PRAISE FOR OTHER TITLES B

The Beginner's Guide to Winning ̶*the Nobel Prize*

He writes like an angel!
Robyn Williams, ABC Science Show

He answers some of the great questions of our age, arguing the case for engaged science, for integrity, creativity and the principle of working for the common good.
Victorian Lifestyle

Topics are embellished with Doherty's wisdom and wit.
Medical Journal of Australia

The Beginner's Guide to Winning the Nobel Prize is an entertaining and insightful book.
Nature Medicine

A Light History of Hot Air

It is a sheer delight.
Outdoor Outlook

Extraordinary outpourings of a brilliant mind.
Launceston Examiner

Doherty has a gift for making the complex appear straightforward … *A Light History of Hot Air* provides a wealth of small pleasures and insights into the workings of the world.
Australian Book Review

A very entertaining and informative memoir and an intriguing observation of human activity.
Australian Bookseller and Publisher

Peter Doherty's pioneering research into human immune systems earned him the Nobel Prize for Physiology and Medicine in 1996. He was Australian of the Year and awarded a Companion of the Order of Australia in 1997 and currently divides his professional time between the University of Melbourne and St Jude Children's Hospital in Memphis, where he is helping unravel the mystery of childhood cancer.

PETER DOHERTY
SENTINEL CHICKENS

WHAT BIRDS TELL US ABOUT OUR HEALTH AND THE WORLD

To Hugh and Irene,
with best wishes,
and thanks for
your help with this.

MELBOURNE
UNIVERSITY
PRESS

MELBOURNE UNIVERSITY PRESS
An imprint of Melbourne University Publishing Limited
187 Grattan Street, Carlton, Victoria 3053, Australia
mup-info@unimelb.edu.au
www.mup.com.au

First published 2012
Text © Peter Doherty, 2012
Design and typography © Melbourne University Publishing Limited, 2012

Melbourne University Publishing acknowledges the kind permission of copyright holders to reproduce illustrations at the beginning of each chapter, particularly Fraser Simpson (pp. 1, 52, 105), Douglas Pratt (p. 9), Mieke Roch (p. 20), canstockphoto (pp. 29, 110, 120), Joern Lehmhus (p. 65), Christopher M. Goldade (p. 86), Keith Hansen (p. 134), Jane Molloy (p. 151) and Phyllis Saroff (p. 164).

Cover design by Design by Committee
Typeset by TypeSkill
Printed by Griffin Press, South Australia

National Library of Australia Cataloguing-in-Publication entry
Doherty, Peter, 1940–

Sentinel chickens: what birds tell us about our health and the world/Peter Doherty.

9780522861105 (pbk.)
9780522861112 (ebook)

Includes bibliographical references and index.

Beneficial birds.
Birds—Environmental enrichment.
Birds—Health aspects.
Chickens—Behavior—Climatic factors.

598.163

CONTENTS

1

Searching for puffins:
an introduction

WITH A BIT OF luck, this was to be the day we would finally meet the puffins. Eight days into our ten-day vacation trip, I still hadn't spotted one. And it wasn't as though they'd been driven away by crowds of other travellers. Our small, 80-passenger boat had been the first of the 2010 summer season to drop down through the Ballard Locks from Seattle's Lake Washington to Puget Sound, before heading north to the San Juan Islands and on past British Columbia to Alaska and our eventual disembarkation in Juneau. We'd pretty much had the protected Inside Passage waterway to ourselves.

With side trips to see glaciers and to visit small communities, it had all been a very enjoyable experience, at least for those whose idea of relaxation is to stand out on a sometimes freezing deck trying to spot wildlife. Every cabin was provided with two sets of powerful binoculars, but there was no guarantee that we would see all the species described in the guidebooks.

So far, though, the sightings of marine and land mammals had more than lived up to expectations. We'd progressively spotted seals, sea lions, orcas, humpback whales, Dall's porpoises, deer, mountain goats, and black and brown (grizzly) bears. Further north, we saw sea otters floating past on their backs with their heads and tails sticking up like black bookends in the current.

Early on, we had sighted a puffin relative, a rhinoceros auklet, and we'd seen other alcids like murrelets and guillemots. Then there were cormorants, terns and gulls of various types. Having come late to bird spotting, much of this was both novel and intriguing. The discovery of a new interest as the years roll on is one of life's good surprises. Of course, I could already recognise many of the familiar bird species. Bald eagles and ravens were everywhere along the Inside Passage. Parked watchfully in convenient locations, they would—like the ubiquitous commuter floatplanes—take off to do loops and bumps along the waterfronts of human habitation. Out on the waterway, we tracked unmistakable white (piebald) eagle heads flying low and fast, while others poked from nests in the tall trees of unspoiled forests. Poised on a calved, floating iceblock, an eagle launched dramatically, then swooped to the glacial sea and soared again, a large fish tightly clasped in its talons.

The dominant, all-seeing eagle has been the symbol of temporal might from ancient Rome, through Byzantium and Czarist Russia to the present-day USA. Wharf-side at the Alaskan fishing port of Petersburg, I had stared, fascinated, at a full-grown eagle perched nearby on a tall post, and soon realised that I was being examined with at least equal intensity. Their binocular eyesight, along with a more sensitive retina and the capacity to see a broader spectrum of colours, gives eagles (and all birds of prey) extraordinarily visual acuity. He saw me much more clearly than I saw him. Still, it's interesting that eagle symbolism is more about all-seeing power than about sanitation, given that the latter may well be the eagle's most important function in nature. Earlier in our Alaskan voyage, we'd watched through binoculars as a few ravens and a host of (largely immature) bald eagles had stripped the corpse of a beached, dead whale.

The iconic bald eagle is, of course, found only in North America. Ravens, on the other hand, are widely distributed across the planet and can be found in every part of the world in which I've lived. In Alaska we were seeing *Corvus corax*, the same species as the famous ravens of the Tower of London, while our backyard in antipodean Australia is sometimes host to the little raven, *Corvus mellori*. (I've mostly just used the common names of birds throughout this book, but a list of Latin binomials is provided at the end.)

Birds are powerful symbols. The indigenous people of this part of North America, the Tlingit, divide themselves into 'eagles' and 'ravens'. We'd seen and heard something of Tlingit culture in the community hall at Metlakatla, a small township that, like many on the Inside Passage, is only accessible by air or water. Traditionally, an eagle has to marry a raven, which is a pretty good way to prevent inbreeding and the genetic diseases that can follow. Eagles and ravens aren't, though, the exclusive avian focus for the Tlingit. They also have a somewhat different, and more pragmatic, relationship with puffins. Taking advantage of available resources, the ancestral Tlingits harvested puffin colonies for meat, skins and eggs.

My obsession with puffins was less utilitarian. With their orange beaks and stocky build, puffins are a particularly endearing bird. Like small children, their heads seem large relative to their body size, which is presumably why Penguin Books branded their kid-oriented series with the amiable puffin. I had lived for a while in Scotland, where puffins are often sighted, but sadly never encountered one. So as our little boat motored quietly towards the bird sanctuary at South Marble Island, I remained alert, eagerly waiting to see my first puffin.

Then, unmistakable because of those bright orange beaks, there were suddenly puffins everywhere, in the air, in the water and in groups on the steep sides of the rocky island where they burrow to make their nests. Wonderful! We'd finally entered the home territory of the tufted puffins. The tufts, or 'tails', of blondish feathers hang symmetrically on either side down the back of the head, differentiating them from the less common horned puffins, which tend to be found further north and are more like the Atlantic puffin that features on the Puffin Books logo.

Watching these tufted puffins up close through the telephoto lens of my camera, I could readily see that, with their relatively short wings and stocky bodies, they aren't such great fliers. It takes a lot for them to launch into the air, and with heads down and wings at maximum flap rate, they strike the water with webbed feet to provide additional momentum. Being both divers and fishers, puffins are very much at home in the sea. The fact that puffins can fly, though, probably protected them from the fate of their much larger, flightless cousin, the great auk, which was hunted to extinction in the nineteenth century.

Even so, puffins haven't been exempt from human harvesting. Apart from helping to feed the Tlingit, they were also once hunted by the Norwegians, who bred a puffin dog (the six-toed lundehund) to dig the birds and their eggs from their burrows. Every species in nature lives by eating one or more other life form, whether plant or animal. It's only when such relationships become unbalanced—as we're seeing now for many of the world's ocean fisheries—that we find ourselves looking at the prospect of starvation and irrevocable species loss.

Humans have sufficient flexibility to change their support systems and switch, for instance, from being traditional fishermen to making souvenirs, becoming tourist guides or engaging in piracy. Birds don't have as many options, and seabird numbers are declining across the planet. Alaska has done a great job of regulating commercial fishing to maintain sustainable populations, which is one of the reasons why we were seeing so many different birds along the Inside Passage. Sensitive and intelligent management has to be our goal for all the world's lakes, waterways and oceans. We must keep the wellbeing of the broader natural world in mind as we seek to ensure our own food supplies. The loss of species diversity affects all of us in myriad ways, from the practical and the scientific to the aesthetic.

* * *

After that final viewing of puffins, sea lions, arctic terns and bald eagles in and around South Marble Island, we sailed on to end our voyage of personal discovery at Juneau, where our little ship abandoned us to depart on its return voyage.

Two days later, we were back in Melbourne and, after struggling through a jet-lagged day, turned on the evening TV news for the first time in about three weeks. Coverage of the massive Gulf of Mexico oil spill that resulted from the explosion of BP's *Deepwater Horizon* rig continued unabated. Both the small screen and the newspapers were showing the usual sad, iconic pictures of oiled pelicans. While conservationists are concerned about the threat posed to the American brown pelican by bacterial diseases like botulism, such hazards are small beer when compared with the impact of a major oil spill. All seabirds are highly vulnerable when there is oil in the water. We weren't in Australia at the time, so we missed seeing the pictures of oiled fairy penguins after the 1995 grounding of the *Iron Baron* in Northern Tasmania, a calamity that involved only 350 tonnes or so of bunker fuel. In Alaska, the vastly bigger 1989 spill that occurred when the *Exxon Valdez* struck Bligh Reef in Prince William Sound is

thought to have resulted in the loss of some 13 000 tufted puffins, though it was their relatives, the pigeon guillemots, that were
most severely affected.

We need to take much more care and to respect the underlying
science, or the increasing degradation of natural habitats that goes
with human population growth and wealthier, high-consumption
lifestyles will inevitably lead to the loss of many bird species.
Close to Melbourne, for instance, there is widespread public concern about the imminent extinction of the orange-bellied parrot.
There have been reports of these parrots being killed by those big,
electricity-generating windmills. Then there's the continued degradation of habitat as we concrete over tracts of land for housing, or
'rehabilitate' wetlands to create up-market golf resorts. Everyone
in the Western Pacific birding community is deeply worried about
what's happening in Asia, as the continuing 'development' of
coastal mudflats—and the resultant loss of whelks, mussels and
the like—depletes the potential food sources for those long-range
migrant species that commute annually between the far north and
the deep south of our small planet.

Though the catastrophic effects of dramatic events like oil
spills are obvious, other issues may not be so familiar unless we
are directly involved. Members of organisations like the Audubon
Society, the Royal Society for the Protection of Birds, and BirdLife
Australia are being recruited to monitor the effects of habitat degradation and climate change on migrating and sedentary bird
species, but this kind of systematic approach attracts minimal
media attention and so mostly goes unnoticed. During the course
of our Alaska vacation, we visited the wonderful Raptor Center at
Sitka and saw the consequences of what happens when powerful
birds fly into power lines or get caught up in discarded fishing gear.
Some of the eagles, hawks and owls that the centre had treated
were so severely damaged that they would never be returned to
the wild. That isn't something I'd thought much about, but now it
comes to mind every time I drive through the countryside or watch

an angler snag and then lose his tackle at the beach. The efforts to minimise the use of plastic bags are more familiar. Floating plastic entangles seabirds. Balled-up and mistaken for food, plastic can choke the birds and their chicks.

Much less in the public eye are aspects of the avian–human interface that are part of my professional world. Trained early on as a veterinary pathologist, I've worked on infection and immunity for almost 50 years, and much of my research over the past 30 or so has been focused on influenza. About four decades back, virologists and epidemiologists began to understand that the influenza A viruses that can be so dangerous to humans are maintained primarily in waterfowl, a finding that has profound implications for human and animal disease. And the massive increase in both the human and domestic chicken population since the mid-twentieth century is influencing the balance between the influenza viruses, wild birds and many mammalian species.

Over the years, I've been privileged to hear some fascinating accounts from people who investigate a range of problems in birds. Being a successful biomedical scientist with an early background in animal health, I'm frequently invited to give talks at veterinary schools. In 2009, for instance, I made my first trip to the excellent South African College of Veterinary Medicine at Onderstepoort, near Pretoria. The Dean, pharmacologist Gerry Swan, told me about the mysterious dying off of Indian vultures and described what he and his colleagues had done to help identify a solution. Then there are the intriguing and little-known tales of how studies in birds and chick embryos have led to massive advances in the understanding of infectious and other human diseases, including cancer.

This book is thus conceived as an exploration of the interactions between the natural world, birds and humans—an exploration that goes beyond the more familiar social and environmental themes in order to discuss a further, darker realm of pathology, poisons and pestilences. Birds have an important monitoring function. Our free-flying, wide-ranging avian relatives serve as sentinels, sampling the health of the air, seas, forests and grasslands that we share with them and with the other complex life forms on this planet. Many bird species migrate globally, so it makes sense for us to know a little of what's happening in both the north and the south of our world. Some of what I relate in the pages that follow will be relatively unfamiliar to even the most committed bird enthusiast. Discussing these stories with my medical scientist friends, I've found that they were equally unaware and fascinated. So this is my best hope: that you will be entertained, informed and, hopefully, even challenged to take action.

2

Distant relatives

WE'VE ALL HEARD ABOUT the coalmine canary that suddenly stops singing and keels over from toxic gas poisoning before the miners are obviously affected. But why is the canary more susceptible? The idea that birds act as sentinels providing us with early warning of potential dangers in the natural world raises some immediate questions: How are avian species similar to mammals like us? And how are they different?

You don't have to be a comparative anatomist, or even the most amateur biologist, to realise that birds are vertebrates and, as such, our distant relatives. But reflecting the demands imposed by an aeronautic lifestyle, the bird skeleton is quite different from the mammalian model. Though birds can be long-lived, with some parrots surviving more than 70 years, they would experience much less back pain than we do, as their lower spinal cords are fused with an extended pelvis. This anatomy helps them to deal with the stresses associated with landing, and gives the structural rigidity necessary to support the powerful movement of muscle, tendons, bones, skin and feathers that enables flight. Also, though humans, birds and some dinosaurs (like *Tyrannosaurus rex*) all share the characteristic of being bipedal (walking on two legs), there is a major difference in the way that this upright stance is achieved.

While our organs, spine and legs align in a vertical plane, the bodies of birds (including the flightless emu and ostrich) and the bipedal dinosaurs are horizontal. As a consequence, birds have more cervical vertebrae (13–25, as opposed to our seven), producing a very flexible neck, which allows the head to swing widely. When on the ground, the bird balances by extending its neck upwards, by using its tail and by bringing the supporting legs to somewhere around the mid-point of the body. As a consequence, the upper leg bone (the femur), which is vertical when we are standing, aligns horizontally along the bird's body and tends to be hidden under the feathers. The top bone that we see when we glance at a bird's leg is the joined tibia and fibula (the lower leg in humans), and below that is the fused tarsus (the equivalent of our foot bones). What looks like the bird 'knee' is in fact the tibiotarsal joint, the equivalent of our ankle. Though penguins may have the appearance of vertically organised humans wearing dinner suits, the way that their bones are aligned is essentially birdlike.

The design of any creature or machine that flies must take serious account of power-to-weight ratios. The wings replace the mammalian forelimbs. And a bird's more porous bones are hollowed out to make for a lighter load. In order to achieve sufficient attachment area for the massive flexor muscles (pectorals) needed to lift a bird off the ground, the clavicles (collarbones) are fused— just like a chicken's 'wishbone'—while the sternum (the bone at the front of the chest) is extended down to produce a deep, vertical 'keel'. Although they are flightless, penguins retain that enlarged sternum to support the muscles they use for swimming, an activity that also goes with having denser bones. Our larger, grounded feathered friends, like the ostrich, emu, cassowary, rhea and kiwi, also have heavy bones, but they are less 'pigeon-chested' as their forelimbs aren't called on to do much work. For that reason, these 'big birds' are classified as ratites, a term that has nothing to do with rodents but is derived from the Latin *ratis*, referring to the flat, 'raft-like' (as opposed to 'keel-like') shape of the sternum.

Though the well-trained, strong flexors of the human leg can drive a pedal-powered 'Gossamer' flying machine, albeit for a relatively short distance, even the most superior athlete lacks both the muscles and the necessary attachment bones to ever flap a lifting wing. Flying cherubim and angels are out in any anatomically realistic world, and the claim that they flew with flapping wings consigns Icarus and his dad to the realms of mythology.

The gravity-defying feat of taking off from level ground or the surface of the sea requires enormous amounts of energy, which in turn means accessing a lot of oxygen (O_2) in the tissues as the body burns glucose to fuel the muscular machine. And the carbon dioxide (CO_2) end product also needs to be discarded. Both mammals and birds have lungs with ever-branching, ever-smaller tubes, which eventually become tiny air capillaries. The barrier between the air capillaries and the blood capillaries is so delicate, and the walls so thin, that the circulating red blood cells (RBCs, or erythrocytes) can easily access fresh air in order to release CO_2 and allow O_2 uptake. But before the air reaches that final, single-cell air–blood interface, there are fundamental differences in the plumbing of the 'pipes' that it travels through. Birds have evolved a more complex, unidirectional, flow-through respiratory system, while the larger and simpler mammalian lung is bidirectional, mixing the inspired (fresh) air with the expired (used) air, even at the level of the terminal, balloon-like alveoli, where the exchange of gases occurs. The bird lung doesn't have 'end structures' like the alveoli, instead using a 'continuous flow' system of tiny, interconnected tubes.

We'll deal with the more familiar human example first. The thoracic cavity that houses the human lung and the heart is separated from the abdominal cavity by the muscular diaphragm. Contracting the diaphragm down expands the chest so that we breathe in, while relaxing the diaphragm allows us to breathe out. The diaphragm doesn't function alone; it is helped along by the intercostal (inter-rib) and abdominal muscles. But as every opera singer knows, it's the diaphragm that plays a key role in driving the expired air past the vocal chords to produce the glorious sound that we recognise as the voice of a Joan Sutherland, a Renée Fleming or a Luciano Pavarotti.

Birds don't have a diaphragm but use other muscles (including the intercostals) to expand or contract the lungs and large air sacs. These air sacs are very extensive and even infiltrate into the bones. Reflecting the nature of that extended, mobile neck, the trachea (the tube that connects the respiratory tract with the outside world) is generally longer in birds than in mammals and acts as a resonator, which is why it's hard for us to mimic some birdcalls. Instead of simply dividing to go to the left or right lung, as it does in mammals, the avian trachea branches out, and these branches pass straight through the lung and terminate in the complex system of anterior and posterior air sacs. No gaseous O_2/CO_2 exchange occurs in the air sacs, which instead function like the bellows that a blacksmith uses to drive air over his forge.

During inspiration, muscles draw the bird sternum forwards and down, which in turn expands the ribs to lower the internal body pressure, bringing fresh air into the lung and air sacs. Half the inspired air passes via the O_2/CO_2 gas-exchange system of the lung to the anterior air sacs, while half remains oxygenated and

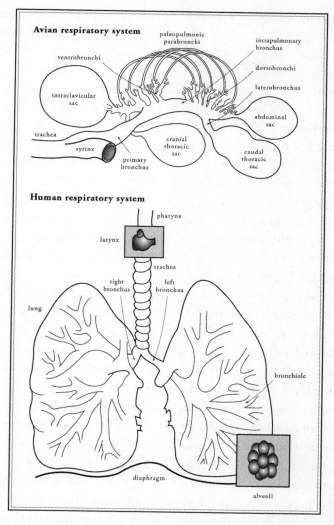

The avian respiratory system is unidirectional, with the O_2/CO_2 gas exchange taking place in the parabronchi. During expiration, air from the posterior sacs moves into the lungs and, simultaneously, air from the anterior sacs moves into the trachea and out of the body. The human respiratory system, by comparison, is bidirectional, with the exchange of gas occurring in those dead-ends of the respiratory tree, the terminal alveoli.

goes directly to the posterior air sacs. Contraction (as the sternum moves back and up) drives the 'spent' air from the lung and the anterior air sacs directly into the trachea, while at the same time forcing fresh air into the lung from the posterior air sacs.

Thus, while our lungs are like tidal systems, with the air from the 'ocean' of the atmosphere washing in and out, the avian strategy provides a continuous stream of fresh air, first from the 'ocean' itself and then from the posterior air sac 'reservoirs' that fill during inspiration and provide fully oxygenated air to the lung during expiration. This continuous-flow bird lung has the additional advantage of assigning the 'bellows' activity to the air sacs, allowing the lung itself to be much more rigid. This means that the thin epithelial layer between the air and the lung capillaries can be organised in such a way that the capillaries themselves can be smaller and the epithelial barrier much thinner, an arrangement that promotes more efficient gaseous exchange.

The net result is that birds do a much better job of accessing O_2 and losing CO_2. Any climber or high mountain skier has watched raptors, geese and the like flying effortlessly at altitudes where low O_2 levels leave us gasping for breath after even minimal exertion. This is also what makes the canary in the mine such a good early warning system. The efficiency of its respiratory system means that it processes a greater volume of air relative to body mass than we do, making it a much more effective 'atmosphere sampling machine'.

One possibility is that this flow-through lung in birds and some dinosaurs may have evolved in response to atmospheric conditions of decreased O_2 and high CO_2 during previous periods of global warming. Mammals would have been very small at that stage, which may have helped them to survive such stressful times. Also, our mammalian ancestors compensated to some extent by taking the design of the RBC a little further. Kicking out the nucleus and other unnecessary elements, the mammalian RBC evolved into a smaller and more completely specialised O_2–uptake and transport unit, able to move through very narrow capillaries. Bird and reptile RBCs remain larger and nucleated.

* * *

It's very obvious that birds and mammals have diverged considerably over the eons, but what do we know about their possible common origin? Way, way back, during the carboniferous period of 350 million or more years ago, the earth was home to small, four-legged, lizard-like creatures, the amniotes, which are thought to have been the ancestors of all birds and mammals. Having crawled out of salt water to live on land, at least some of the newly terrestrial vertebrates used strategies for sexual reproduction that allowed them to move further afield and to exploit drier and more unpredictable environments. Perhaps the amniotes evolved directly from a viviparous marine species that, like the shark, carries and nurtures the developing young within the body of the adult female, rather than from that ultimate uninvolved mother, the fishy female who first generates then releases enormous numbers of eggs into ocean or river waters to be impregnated by a cloud of sperm provided by an enthusiastic though equally detached male.

The newly evolving amniotes of the very deep past enjoyed a much more intimate interaction as a precursor to producing substantially smaller numbers of nurtured offspring. Though all male mammals have a phallus, that's true for only a few species of birds, particularly the big, earthbound ratites. Those birds that lack this structure simply oppose their common urogenital openings (the cloaca) to allow sperm to pass from the male to the female. Among the mammals, the viviparous marsupials and placentals (including us) give birth to live young, while the small group of monotremes (the platypus and the echidnas) are oviparous—that is, like birds, they lay eggs. Birds are never viviparous. Most reptiles are also oviparous, and the females of different species retain their eggs within the body for varying periods, though some do nurture and give birth to live offspring. All species of mammals (including the monotremes) at first feed their young from a specialised, lactating mammary gland. A few bird species, including pigeons, doves, flamingos and some penguins, secrete high-protein, high-fat 'crop milk', which has a cottage cheese–like consistency. Both males and females make crop milk, so while the placental mother is inevitably tied for a time to her developing offspring, and the marsupial mum carries her infant in a pouch, nurturing the young is a role that is passed on to the male in a number of bird species.

Whether within the protective shell of a bird's egg or in the human uterus, the developing amniote baby is surrounded and sustained by membranes—the amnion, chorion and allantois—that are produced from the tissues of the embryo. In humans, the chorion invades the endometrium of the mother's uterus, forming the placenta, which allows the transfer of nutrients and allows the O_2/CO_2 exchange that is essential for life. The vitamin and mineral nutrient store for the growing bird is provided in the lipid rich, bright yellow yolk, and the proteinaceous albumen of the egg white. The fused chorioallantoic membrane that lines the porous shell provides the necessary interface between the fetal blood supply and the surrounding atmosphere. In both cases, the embryo develops within the fluid-filled amniotic cavity. The imminent arrival of a human baby is announced by the waters breaking, the rupturing of the amniotic sac. We know that the baby bird is about to emerge when we hear the pip-pipping sound as its beak breaks through the surrounding shell.

* * *

Birds and mammals were well and truly distinct by the time that large dinosaurs disappeared during the late Cretaceous and early Eocene more than 50 million years ago. Our world, which is about 4.6 billion years old, was a very different place at that time. The global climate was much warmer than it is today, and atmospheric CO_2 levels were at least twice as high (750 parts per million, or ppm) as they are now. Apart from the blue and grey of the oceans, the planet was green from pole to pole. The Eocene and the Oligocene that followed also saw the final fragmentation of the ancient mega-continent of Gondwanaland. This happened very slowly in the biological sense, allowing plenty of time for animal and plant species to evolve and adapt to the inevitable process of change.

Many scientists believe that the birds are the living descendants of the dinosaurs. Using the latest sequencing techniques to analyse small fragments of protein recovered from a 68-million-year-old fossil of *Tyrannosaurus rex*, scientists concluded that birds are more closely related to that giant predator than to the reptiles of today. Then there's the suggestion, based on the analysis of dinosaur bones, that dinosaurs also had air sacs, and the fact that some dinosaurs, like many birds, had a muscular gizzard and swallowed stones to help grind food. Of course, birds dispensed totally with teeth and developed a beak instead, which is not great for grinding. Some theropod dinosaurs (*T. rex* is a theropod) had hollow bones, while recent archaeological evidence from China supports the idea that at least one member of the family had feathers. Whether the toothy, flying dinosaur *Archaeopteryx* is in the direct line of bird ancestry is still debated.

Do those flightless big birds, the ratites—the ostrich, emu, cassowary, rhea and (now extinct) moa—inform us about the depth of avian antiquity? If, indeed, these natives of Africa, Australia, New Guinea, South America and New Zealand are all descended from a common, grounded ancestor species, then the ratite Adam and Eve must have been around at the time Gondwana started to break up approximately 169 million years back. It's tempting to conclude that the African ostrich, Australasian emu and cassowary, and New Zealand moa all evolved further to become distinct species after being isolated on unconnected land masses. Certainly this theory seems to explain the

great similarities between the different big birds, but science offers
a different and somewhat counter-intuitive possibility. There's no
evidence that the ostriches, emus and so forth are in any way
closely related to those great swimmers, the penguins, which do
move readily between land masses, but different ratites might
just possibly have developed more recently from an earlier, flying
version of the distantly related tinamids. The implication would
then be that the similarity between these big birds reflects conver-
gent evolution, as different lineages living on isolated continents
developed similar characteristics in response to comparable selection pressures. In the not too distant future, this question will no
doubt be resolved by DNA analysis.

Our transition to an erect, bipedal lifestyle is much more
recent and better understood. Big-brained *Homo sapiens sapiens*
(as distinct from *Homo sapiens neanderthalensis*) has only been
around for 120 000–200 000 years at most. There's no possibility
that our distant relatives, like *Homo ergaster* and *Homo erectus*
from maybe a million years back, could ever have flown between
continents. In fact, DNA analysis tells us very clearly that we are
all 'out of Africa' and that, if we confine our focus to long-term
indigenous populations, the extent of genetic diversity within any
particular human group decreases the further we get from Africa.
Birds colonised this planet long before we did, and even if their
ancestors were the ultimate in scary dinosaurs, they can never
have been as frightening, dangerous or damaging to other life
forms as we are today.

Pointing to the pattern of global domination followed by
extinction that we associate with animals like *T. rex*, some people
suggest that humans are behaving like dinosaurs when we fail
to act in ways that minimise the damage we are doing to this
small, green planet. That's unfair to the dinosaurs. While our
wilful behaviour may ultimately ensure our elimination, and
that of many other life forms, there is no way that the big dinosaurs could have deliberately contributed to their own extinction.

Consciousness, reason, callous greed, deliberate ignorance and true malevolence are very much unique to *Homo sapiens sapiens*.

There's also a tendency to describe those who will neither acknowledge nor embrace the need for action on issues like anthropogenic climate change as 'ostriches', alluding to the myth that this big bird just buries its head in the sand whenever a threat appears on the horizon. Different explanations have been suggested for this odd behaviour, including the possibility that a nesting ostrich digs out a hollow to accommodate its head and be less conspicuous, and the alternative idea that the ostrich scrapes its head along the ground to pick up stones for its grinding gizzard. So maybe its behaviour is not so absurd after all. Perhaps we would do better to keep our own heads that close to the earth so that we could look more clearly at what we are doing to it.

3

Chick embryos and other developing life forms

EGGS IN ALL THEIR culinary forms—poached, boiled, scrambled or fried—are pretty familiar to most Westerners, but in this era of industrialised and highly regulated food supply, few encounter a chick embryo over breakfast. In my twenties, though, fertilised hen's eggs did become an unusually familiar part of my life, but for reasons that were only indirectly related to food.

When I enrolled as a 17-year-old undergraduate in the University of Queensland School of Veterinary Science, I had absolutely no interest in the problems of chickens or, indeed, of any other bird. I was not typical of most veterinary students in that my aim from the outset was to become a research scientist studying the diseases of large domestic animals, particularly the cattle and sheep that were then the mainstay of the Australian economy. Many of my predominantly male fellow students had rural backgrounds and were also intending to work mainly with food- and wool-producing ruminants. A few were headed directly for careers in the horse racing industry, and some would locate in the city or suburbs to pursue what was then referred to as 'dog and cat' practice.

I don't recall any fellow student who was dedicated to the idea of working with poultry ('chooks' in the Australian argot) or with birds in any form. Large commercial poultry operations were still to emerge on the local scene, and the rare 'chook man'— as avian disease specialists were dismissively called—was usually employed in government service. Apart from providing advice to poultry producers, they also had a regulatory role in monitoring and then dealing with outbreaks of infectious diseases. It's hardly surprising that someone with the power to order that a virus-infected flock be killed, burnt and buried should be treated with a degree of trepidation by farmers, even if eliminating the pathogen is ultimately to their benefit.

So birds were pretty low on the totem pole when it came to the veterinary school agenda, and over the course of the five-year degree, we spent relatively little time on avian pathology or medicine, although domestic chickens and their eggs did receive some attention from the microbiology and public health aspect. Infection can devastate poultry flocks, and eggs loaded with *Salmonella* bacteria are a frequent cause of food poisoning.

It was in my second year at university, though, that I discovered what still rates in my mind as one of the great books on basic biology, Bradley M Patten's 1920 oeuvre, *The Early Embryology of the Chick*, which introduced me to the wonders of the chick embryo. Bored, broke and at a loose end on one of the short university vacations, I read and re-read Patten's little treatise. Tracing the steps that take a fertilised ovum via a disc of vascularised cells in the yolk sac to the 'primitive streak', then the mesoderm, endoderm and ectoderm, which differentiate progressively to build the complexities of a multicellular, multi-organ vertebrate is a particularly easy process to follow visually for the developing chick embryo.

Unknowingly, I was following the path to biological enlightenment that Marcello Malpighi first took in seventeenth-century Bologna. After all, it's just a matter of assembling a bunch of eggs that are all fertilised at the same time, then breaking them open at regular intervals and laying out the contents on a flat surface. That's not a difficult anatomical preparation, and even in late medieval times, probing the mysteries of hens' eggs posed no threat to the fanciful doctrines of 'revealed' religion.

*** *** ***

After graduation, I was employed as a veterinary pathologist at the Queensland Department of Primary Industries' Animal Research Institute, which served as a diagnostic laboratory for the network of government veterinary officers and stock inspectors. That's where I met petite, blonde and pretty Penny Stephens, who at the age of 21, and just graduated from the university's microbiology department, was the institute's first full-time animal virologist. We married two years later and are still together.

That early post-mortem room experience also failed to feature birds, as most of the avian disease problems were looked after more by the microbiologists than by the pathologists. Penny, however, was working on avian infectious bronchitis virus (IBV), a respiratory infection caused by a coronavirus. IBV doesn't infect humans, but as it diminishes both weight gain and egg production, the poultry industry vaccinates to prevent this cause of economic loss. Way back then, the disease had only recently been recognised in Australia, and Penny was studying one of the first IBV isolates by injecting the virus directly into fertilised hens' eggs. As IBV grows, it kills the baby chick, so when the eggs are opened

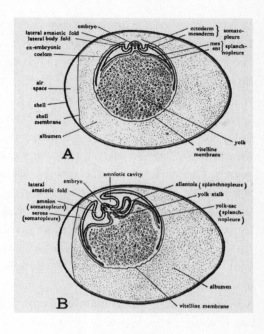

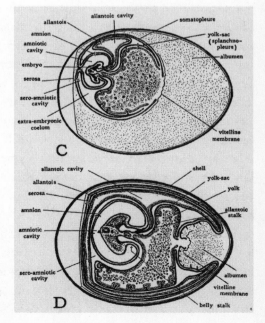

Figure from Bradley M Patten's *The Early Embryology of the Chick*, showing the chick embryo at about one, two, five and 14 days of incubation.

at term, the embryo appears to be dwarfed. That may seem a bit bizarre, injecting a suspected pathogen into embryonated eggs, but Penny was actually following a long and very successful tradition in animal virology.

* * *

From the 1930s through to the 1950s, much of the serious research done on viruses that infect vertebrates involved chick embryos somewhere along the line. Viruses can only reproduce within living cells, and though their 'blueprint or barcode' is written in the nucleic acid (RNA or DNA) sequences that they carry with them from, say, birds to mosquitoes to humans, they also exploit the 'machinery' of the various host cells they infect to replicate and survive.

That's where chick embryos come in: it just so happens that many of the viruses that infect us will also grow in embryonated hens' eggs. So virology laboratories of the past featured large, industrial egg incubators. The fertilised hens' eggs are purchased from commercial suppliers and transported without harm in unheated cartons. The embryos don't continue to develop until they find themselves at around 38°C, about 3°C lower than the normal bird body temperature. On arrival at the laboratory, the eggs are first 'candled' to see if they contain a live fetus. The shell is sufficiently translucent to pass light from a 20-watt bulb or the like, and the red spot of the fertilised disc on the yolk sac is readily seen. The embryonated eggs are then arrayed in rows on the shelves of the incubator, which has some form of mechanism for turning them at regular intervals, a job that is done in nature by the nurturing bird parent.

This technique of injecting suspected virus-containing material into hens' eggs—archaic though it might seem to those of us who are accustomed to the TV stereotype of a white-coated scientist working in an aseptic laboratory full of gleaming instruments and machines—was very important until the 1960s and is used even now for the production of most influenza vaccines. At the stage that I first learnt diagnostic microbiology almost 50 years ago, the classical method for isolating then growing *Chlamydia* involved injecting suspected material straight into the yolk sac of embryonated hens' eggs. Bacteria in this family cause psittacosis (*Chlamydia psittaci*, a respiratory infection of parrots that can transmit to their owners), the human eye disease trachoma (*Chlamydia trachomatis*) and abortion in sheep, cattle and, occasionally, humans (*Chlamydia abortus*). The trained eye soon learns to spot the red, spore-like *Chlamydia* elementary bodies in Giemsa dye-stained yolk-sac smears. Inoculation directly into the yolk sac was also used to grow and isolate some of the rickettsia, including the bug that causes Q fever (*Coxiella burnetii*), a common problem for abattoir workers that has now been minimised by the development of a protective vaccine.

FM (Mac) Burnet was the great proponent of infectious disease research with chick embryos. A virologist by training, he had first worked with bacteriophages (viruses that infect bacteria) while completing his graduate work in London. Back in the 1930s, this area of research focused particularly on 'phages' growing in bacterial 'lawns' maintained in flat, round, glass Petri dishes containing some form of 'nutrient agar'. Each 'spot' on a contiguous bacterial 'lawn' represents a single 'clone' of virus. The invading virus particle first kills the cell that it has parasitised, then its progeny progressively infect other bacteria nearby to give a clear, spreading 'plaque' of virus growth. For researchers like Burnet, this technique had three great advantages. The first was that they could 'pick' the individual plaques using a rubber bulb (for gentle suction) on the other end of a sterile glass pipette, allowing these 'cloned' viruses to be further cultured and analysed. The second was that any change to a 'less fit' state (diminished growth giving smaller plaque size) allowed the visual recognition then isolation of mutants for genetic studies. The third was that the number of viruses in the input inoculum could be measured by just counting the plaques—and good science is all about accurate measurement.

Late in the 1930s, wanting to use the type of approach that he'd used with the phages to study viruses that infect humans, Burnet picked up on a report from Ernest Goodpasture and Alice Woodruff at Vanderbilt University, in Nashville, Tennessee. Goodpasture and Woodruff had developed a spectrum of different chick embryo inoculation techniques for working with infectious agents, but what particularly fascinated Burnet was their procedure for 'dropping' the chick embryo chorioallantois, the fetal membrane that is apposed to the inner layer of the porous (to air) egg shell and allows the embryonic red cells to 'breathe' oxygen directly from the atmosphere.

The procedure that the virologists used for chorioallantoic membrane inoculation involved wiping the shell with alcohol to kill any bacteria or fungi, then using a little grinding wheel spinning in the handset of an old-style dental drill to make two small holes, one over the air space at the more rounded end of the egg, and the other midway along the more horizontal aspect. Care was taken not to go through the tough, white, non-living fibrous layer that's on the inner aspect of the shell, which was then breached minutely at both sites with a previously sterilised sharp needle. Taking a rubber pipette bulb, suction was gently applied over the air space. As that air was drawn out, the change in pressure caused the chorioallantoic membrane to drop away from the long side of the egg, giving a flat 'lawn' of living chick cells that could be inoculated from above (through the same hole) with the organism of interest. The openings were then sealed with drops of hot paraffin wax, and the injected eggs were incubated for several days at 37–38°C to allow the growth of discrete virus colonies, just like those analysed for bacteriophages grown on plated bacteria. This procedure was used with great effect by Burnet (and others) to clone, isolate mutants and count viruses like influenza, the

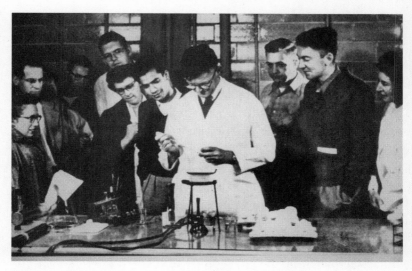

FM Burnet demonstrating his egg inoculation technique.

various poxes (including smallpox) and the herpesviruses. People still use the method occasionally today to grow cancer cells.

Burnet took the understanding of genetics, mutation and clonal lineages that he developed through his bacteriophage, vertebrate virus and chick embryo research and applied it conceptually to the emerging field of immunology, resulting in a theory that became the central dogma of the subject: 'The clonal selection theory of acquired immunity'. His Nobel Prize could well have been for clonal selection. Both he and many others thought it was his most valuable contribution, though he would have been paired with David Talmage from Denver, who came to the same conclusion from a different intellectual background. The Nobel that he did share with Peter Medawar was for the 'Theory of immunological tolerance'. Burnet continued to do hands-on science for about five years after he was honoured by the Swedes, when he hit the then compulsory retiring age of 65 and left the laboratory to pursue his other great passion, writing conceptual books about science. Some of his last experiments used the chick embryo chorioallantoic membrane to explore questions about tissue graft rejection and immune tolerance.

Sir Macfarlane Burnet chatted with Ghandi, met Kings and Queens, and received the highest orders of the old British Empire. He gave every prestigious lecture and was generally applauded as a very great man, an opinion that was evidently agreeable to him. On the other hand, he stayed honest and kept on with his 'day job' of drilling holes in eggs and dropping stuff onto chick embryo membranes. One of the best photographs of Burnet is in a white coat, demonstrating his egg inoculation techniques to students and faculty at the University of Wisconsin. There we have the public and the private face of science, the perception and the practice. Staying grounded in reality is everything, and some of our current political leaders might learn from Burnett's example. But maybe such groundedness belongs to a past era— an era when chick embryos were at the cutting edge of infectious disease research.

4

Sentinel chickens

THE IDEA OF 'SENTINEL chickens' seemed pretty incongruous when I first heard the phrase as a young undergraduate. My reaction was no doubt conditioned by recollections of the scatty and fussy hens that scratched about in the dusty chicken run in my grandmother's backyard. The notion of the humble chicken waiting like a trained soldier, alert and focused, for some unseen and approaching enemy just didn't seem likely. Hens *en garde*!

Like most students in that distant era, I knew everything and knew nothing. Nowadays, any reasonably sophisticated young person would go immediately to the internet and find that way back to mythological times, guard duty has been part of the avian job description. Gods with the body of a man and the head of a bird, like the ibis, falcon, hawk or heron, watched over the ancient Egyptians. In the Western tradition, the cockerel, or rooster, symbolises vigilance and has been widely used as a French heraldic device. Adopted as the national symbol at the time of the 1789 revolution, the proud, colourful rooster of France (*le Coq Gaulois*) went beak to beak with the black eagle of Germany during World War I.

When it comes to warning us of imminent danger, sentinel geese have long been associated with the human story. Geese go on the attack and make an enormous noise if they perceive an incursion into what they regard as their patch. The trick is to provide feed and nurture so that they make our patch their patch. According to the Roman historian Titus Livius—better known as Livy—sacred geese in the temple of the Goddess Juno alerted the exhausted defenders of ancient Rome to a nocturnal attack by marauding Gauls. In modern times, Scottish whiskey distilleries are sometimes guarded by gaggles of geese that raise a loud hue and cry if a thief tries to make off with what many consider the most spiritual of all aqua vitae. Whiskey may be part of the local religion, but I doubt whether the pragmatic Scots would regard the birds as sacred.

Then there's the story of the ravens that somehow guarantee the integrity of the Tower of London and, beyond that, the continuity of the crown of England. Legend has it that the monarchy will fall when the ravens leave the tower. Following the spirit of a decree by Charles II, there are always at least ten ravens available, six on duty and four active reservists. Cared for by a raven master drawn from the ranks of the Beefeaters, the medievally attired tower guards, clipping one wing ensures that the resident ravens can't fly away.

* * *

I first heard the term *sentinel chicken* from an older cousin, Ralph Doherty, a medical scientist who was then building a substantial reputation in the study of the insect- or arthropod-transmitted

viruses, known collectively as the arboviruses, also called toga-
viruses because they have an outer 'envelope' or 'coat'. Among
the major achievements of his research group at Brisbane's
Queensland Institute of Medical Research was the discovery that
the mosquito-borne Ross River virus (RRV) is the cause of the
human disease epidemic polyarthritis with rash, a painful and
debilitating condition that can persist for several months. With
more than 4000 cases every year, this non-fatal disease is all too
familiar to those who live in the northern parts of Australia and
has been rapidly spreading away from the tropics.

Like all viruses, the arboviruses can only reproduce themselves within living cells. What makes the arboviruses special is
that they replicate in the tissues of very different types of animals,
though the individual viruses in this very large group do vary
considerably in their overall host range. The 'virus production
factories' include biting insects, particularly mosquitoes and ticks,
which, as they take their blood feed, either become infected or
(if already carrying the virus) transmit the infection to warm-
blooded species, including human beings and a whole spectrum
of furry and feathered vertebrates.

And that's why we have sentinel chickens. The progressive
spread of many arboviruses is monitored by placing caged chickens
around the countryside at sites where they are likely to be bitten
by mosquitoes. The widely distributed birds are sampled regularly, a comparatively non-intrusive process that involves taking a
small amount of blood from the prominent wing vein. The blood
is allowed to clot, and the yellowish serum supernatant is either
frozen or taken on ice to a specialist laboratory, where the samples are analysed for seroconversion. That is, the technician uses a
well-established assay to detect newly acquired (since the previous
test) circulating antibodies specific for the virus in question. (That
doesn't work for all mosquito-borne infections. Chickens aren't
very susceptible to RRV, for example, which seems to prefer mammalian hosts, and they're of no value for tracking malaria, for

which we humans are the most sensitive sentinels. As explained in Chapter 11, birds have their own distinct malaria parasites.)

If, for example, the birds were seronegative when taken to their guard station, then seropositive for some arbovirus six months later, it's obvious that they were exposed to an infected mosquito at some time over that period. The relatively few virus particles injected by the feeding mosquito will have travelled via the circulation to invade susceptible cells in one or other organ of the new chicken host. Successive cycles of virus replication then lead to the presence of a great deal more virus in blood (viremia), a process that terminates somewhere over the next 7–12 days or so, when the developing immune response will lead to the production of specific, neutralising antibodies. Those antibodies will continue to be made for the life of the bird. Once antibody-positive for the infection of interest, the chicken veteran is both permanently immune and eligible for honourable retirement and replacement with a new recruit.

* * *

Virologists further sub-classify the arboviruses into alphaviruses and flaviviruses. The alphaviruses include RRV and Barmah Forest virus (in Australia), eastern equine encephalitis virus (in the USA) and the Chikungunya virus that has lately been spreading from the Indian Ocean region to South-East Asia and the Mediterranean. Human infection with Chikungunya, RRV or Barmah Forest virus can lead to the development of persistent polyarthritis with rash, while chickens, at least, remain asymptomatic.

All the flaviviruses are broadly related to yellow fever virus (YFV), the terrible pathogen that kills humans by a combination of haemorrhagic disease and liver destruction. That's where the 'flavi' (Latin for yellow) comes from, describing the severe jaundice that characterises the lethally compromised patient. The main vector is the mosquito *Aedes aegypti*, which is present in tropical North Queensland, though there have been no cases of yellow fever in Australia. A vaccine was developed in the 1930s by the South African medical scientist Max Theiler, an achievement recognised by his 1951 Nobel Prize. There are, however, 70 known flaviviruses, with 30 of these being found in southern Asia and the Australasian region. Some are 'orphan' viruses that are not associated with any known disease.

Way back in the 1960s, cousin Ralph's involvement with sentinel chickens reflected the broad interests of his research group in arbovirus epidemiology, the study of how this diversity of infections spreads and is maintained in nature. Some arboviruses, particularly the tick-borne ones, can 'overwinter' by vertical transmission through the successive stages of an insect life cycle, but even when this does occur, it's unlikely to be the main mechanism that keeps the virus going in nature. Though infectious disease epidemiologists search for the vertebrate 'maintaining hosts' that continue the mosquito–animal transmission cycle, the identity of the key species can be incredibly hard to nail down. Antibodies (the footprints of prior infection) to RRV have, for example, been found in marsupial and placental mammals and, less often, in birds, but that doesn't prove that the levels of virus in blood were sufficient to cause widespread infection of the mosquito vectors. This two-way insect–vertebrate interchange probably continues throughout the year in the warmer parts of a continental landmass, particularly in forested areas where there is no effective mosquito control. Migrating birds are, of course, likely culprits for any north or south spread away from the tropics with the onset of spring and summer.

* * *

Over the years, one of the medically important functions of Australia's valiant sentinel chickens has been to serve as 'birds of record' for measuring the southern spread of Murray Valley encephalitis virus (MVE), a flavivirus that's also called Australia encephalitis virus. This infection becomes a problem when the combination of warm weather and an abnormally wet season leads to a massive increase in mosquito numbers. If MVE is somewhere in the neighbourhood, perhaps at high enough levels in the blood of susceptible birds, then mosquitoes become infected and sporadic cases of encephalitis are seen in humans, particularly those living along the banks of major water courses like the Murray River. Though MVE has also been found in Papua New Guinea and Indonesia, the main threat to our north is the closely related, but much more dangerous, Japanese encephalitis virus (JEV), which causes severe disease in a relatively high proportion of infected people. Pigs, rather than birds, are known to be a major maintaining host for JEV, and one way of protecting humans is to decrease the 'multiplier' factor by vaccinating pigs. There are also effective human vaccines for JEV. According to public health doctors, JEV is not a cause of locally acquired disease in the USA, perhaps because of the lack of the main vector, *Culex tritaeniorhynchus*. This mosquito is also absent from Australia, but an alternative vector, *Culex gelidus*, has been identified in the tropical north, where there have been two fatal JEV cases.

Staffed by successive generations of avian 'volunteers', at least some of those sentinel chicken outposts that were located around the country to inform us about the spread of MVE in the 1960s and 1970s still house birds on active duty as part of a continuing Australian surveillance network. Sentinels in the cooler south seroconvert to MVE from time to time, though most evidence of infection is found in tropical northern Australia where occasional human outbreaks continue to occur. The Australian chickens also pick up evidence for the circulation of the closely related (to MVE) Kunjin virus, an occasional cause of human encephalitis, and Barmah Forest virus. Kunjin recently (2011) caused a number of deaths in Australian horses, and is very closely related to the West Nile virus discussed in the next chapter.

The use of sentinels depends, of course, on knowing the identity of the virus that's being looked for. Otherwise, it isn't possible to set up a specific antibody test to determine if any individual—whether poultry or person—has indeed been infected. Though human outbreaks of what was then called Australian X disease had been recognised as early as 1917, it wasn't till 1951 that Eric French, then working at the Walter and Eliza Hall Institute in Melbourne, reported the isolation and initial characterisation of the MVE virus (see Chapter 9).

Apart from the information from sentinel chickens and human cases, what else is known about MVE? The mosquito vector, *Culex annulirostris*, has been identified, but there are only indirect antibody results that implicate several species of cormorants and the Nankeen night heron as possible maintaining hosts. The Nankeen night heron is common in the wetter regions of southern and northern Australia and is generally regarded as a non-threatened species. It does depend heavily on access to fresh water, and there was some cause for concern during the recent long drought, now broken by the return of an unprecedented La Niña climate system, bringing severe flooding and massive cyclonic activity. That, of course, is also likely to increase the incidence of mosquito-borne infections.

Sometimes it's a relatively straightforward matter to establish that a particular species of bird is susceptible to a given arbovirus infection and is capable of circulating the virus, thereby functioning as a maintaining host. For example, eastern equine encephalitis virus, an alphavirus that circulates in the USA and causes disease in both horses and humans, also kills significant numbers of ibises, starlings and emus. Both the birds that eventually die and the survivors can have very high levels of virus in blood.

In general, though, it's been easier to identify the insect vectors that transmit these infections than to establish which particular wild birds or mammals support their overwintering. One reason for this is that arboviruses generally persist longer in mosquitoes, as they lack the type of adaptive, or highly specific, immune system that is characteristic of birds, mammals and the other bony vertebrates. Even when vertebrates suffer a severe infection, the virus is usually eliminated from the blood of survivors within 8–12 days. The other reason is that trapping and handling wild birds takes a lot of effort, while it's relatively easy to catch large numbers of mosquitoes using light traps that emit CO_2 and other chemical attractants (like octenol), simulating the presence of warm-blooded animals. A more primitive technique is to allow them to bite, say, a tethered horse or your own arm, then capture them using some sort of suction device that may be as simple as a skilfully used drinking straw.

Once trapped, the mosquitoes are classified by a medical entomologist, then those of the same type are pooled, frozen and later ground up in saline for injection into some detection system (such as tissue culture or suckling mouse brain), which will then grow any virus the mosquitoes were carrying. The freshly isolated viruses can then be identified by sequencing to determine their characteristic genetic code, using essentially the same technique that forensic experts employ to identify DNA from a rapist or murderer.

* * *

The capacity to produce highly specific antibodies following natural infection in the field or forest is the basis of the sentinel chicken's role. We feed and nurture these doughty guardians because birds have both a thymus that produces the immune T lymphocytes (including the killer T cells, which I've worked on for almost four decades) and the B lymphocytes or plasma cells that produce the specific antibodies we detect in blood. As discussed in more detail in Chapter 14, the avian and mammalian immune systems have evolved somewhat differently over the eons, but they do the same job of controlling infection. Furthermore, this shared capacity for generating long-term immune memory is the reason why, in the past century or so, we have seen the development of numerous protective vaccines for both domestic birds and chickens.

Arboviruses aren't, however, on the chicken vaccine list, as they don't affect commercial producers. How vaccines are used is always determined by practical considerations, and the fact that a product is used in one vertebrate but not another doesn't reflect some sort of discriminatory 'speciesism'. For obvious reasons, it's pretty much impossible to vaccinate wild birds against anything. In the USA, valuable horses are vaccinated against the Venezuelan equine encephalitis alphavirus, while humans are not. People who live in the more prosperous countries are protected by the environmental control of mosquitoes that's practised in most of the larger, warmer cities, by a more indoor lifestyle and by the judicious application of mosquito repellent when venturing into the countryside. We've never made a vaccine against MVE because the incidence is too low, but such a vaccine could be developed if, for example, the warming associated with anthropogenic climate change led to MVE becoming a more substantial threat to large numbers of humans.

A more likely danger for Australians is that infections like JEV and malaria will simply migrate south as ambient temperatures rise, birds modify their migration patterns, and mosquitoes extend their host range. That is already happening in parts of Africa, as infected mosquito populations move inexorably into the cooler and higher regions of the continent, which were formerly malaria-free. In Europe, Chikungunya virus has now penetrated as far north as Ravenna. As land, air and water temperatures increase, the shift of viruses that depend on a mosquito–vertebrate (bird or mammal) lifecycle into what were temperate regions will inevitably continue.

5

Falling crows

TALKING CASUALLY WITH AN immunologist colleague over coffee, the conversation wandered to the delights of golfing on a beautiful course overlooking the sea. Somewhere in California, as I recall—maybe Torrey Pines at La Jolla?

'An odd thing,' he says. 'Hitting into the rough, I came face to face with a cage full of chickens! Chickens on a golf course! What's that about?'

'Hah, sentinel chickens,' I replied. 'They've been placed there to monitor the spread of WNV, West Nile virus.'

My scientific world is largely split between two disciplines, virology and immunology. Any virologist would have known immediately what those birds were for. The fact that my

immunologist golfing friend hadn't made the chicken–WNV connection reflects the ever-increasing depth of specialisation in the
sciences and the enormous expansion in knowledge. We working
scientists increasingly find ourselves living in a kind of Tower of
Babel, where it's harder and harder to stay abreast of what's going
on in even closely related fields. In the research and writing of
this book about the interface between birds, humans, diseases
and environmental degradation, I've been forced to confront my
own colossal ignorance of a number of related areas, particularly
zoology and ornithology.

But I do know something about West Nile virus—indeed I
have worked with similar pathogens. Everyone with any interest
in the epidemiology of infectious disease is very aware of what
began in New York City in the summer and fall of 1999. Crows
fell from the sky; a whole spectrum of exotic and native birds
(including Chilean flamingos and bald eagles) died at the Bronx
and Queens zoos, and some 62 people developed neurological
symptoms, with seven deaths. First isolated in Uganda in 1937,
WNV is a flavivirus that was a known cause of periodic encephalitis and meningitis outbreaks in human populations in parts
of the Old World. The invading New York WNV strain, isolated from one of the dead flamingos, was identified by genetic
sequencing as being essentially identical to a virus that infected at
least 500 people in Bucharest in 1996, caused some 400 cases in
Israel in 1998 and was responsible for more than 40 fatalities
in the Volgograd region of Russia in 1999.

Classified as a member of the Japanese encephalitis group that
also includes the Australian Kunjin virus, the New York WNV
epidemic marked the first time that this virus had caused problems in North America, though it turns out that there had been
earlier cases of human WNV infection on Manhattan Island. Back
at the beginning of the 1950s, Chester Southam and Alice Moore
injected a spectrum of viruses, including WNV, into people with
inoperable cancer. The hope was that these viruses, which killed

similar cells in tissue culture, might grow selectively in, and eliminate, the rapidly dividing tumours. However, though they were
able to show that WNV did indeed cause asymptomatic infections in five of their 21 patients at New York's Memorial Cancer
Center, there was no suggestion of any clinical benefit. Today, of
course, nobody would even think of deliberately injecting very
sick people with potentially lethal viruses. There were no ethics
committees to monitor such activities in those distant days of
human experimentation.

In fairness to Southam and Moore, though, it is important to
recognise that they acted in good faith at a time when we knew so
much less about infectious diseases and cancer. Also, their small
study has ultimately proven to be of value, as it established definitively that the level and duration of WNV circulation in human
blood can be sufficient to infect biting mosquitoes. Perhaps the
1999 WNV outbreak strain crossed the Atlantic in a viremic traveller who was incubating the disease. Only about 30% of those
infected with WNV develop noticeable symptoms, and younger
people, who are more likely to be adventurous in ways that might
bring them into contact with infected mosquitoes, are much less
likely to be clinically affected.

Then there's also the possibility that WNV traversed the
ocean barrier in imported birds. Either way, exposure to local
mosquitoes on the ground at JFK or Newark airports could have
led to that initial transmission and, ultimately, to the establish-
ment of WNV as a cause of endemic infection in the Americas. As
with many situations where a species suddenly 'breaks out' and
spreads more widely, the outcome might simply be the sum of a
number of essentially random events.

Both the distance across the Atlantic to New York City and
the lethality of WNV for wild birds would seem to minimise any
possibility that the virus was carried from Europe or Africa during
some normal process of avian migration, though not all species

are as susceptible to this disease as American crows. And there's always the remote possibility of a laboratory escape, but it seems extremely unlikely that a mosquito could have taken a blood meal from an infected vertebrate, perhaps a researcher or a laboratory mouse, and then carried WNV out into the wild. Anyone working with such a virulent virus would be extremely careful, and all laboratory animals, especially those infected with exotic pathogens, are kept in rigorously inspected, high-security facilities that certainly exclude mosquitoes.

Bioterrorism is also a possibility, but no group claimed responsibility. While there was some 'weaponisation' of such viruses during the Cold War, they were never considered to be a very practicable component of any military arsenal. In this case, weaponisation simply meant making lots of the stuff to spray at some unfortunate enemy. Laboratory infections caused by accidental exposure to aerosols containing high virus concentrations are known to occur, but free-floating arboviruses are too labile to spread far through the air. The USA did, though, grow vast numbers of yellow fever virus–infected *Aedes aegypti* mosquitoes at their Pine Bluff Arsenal before we finally regained our senses and such dangerous idiocy was banned, along with everything else of that type, under the 1975 Biological Weapons Convention. Any US stockpiles had been destroyed by the early 1970s.

A prominent characteristic of the New York and other recent WNV outbreaks has been very high mortality rates in some wild bird species. One possibility is that this reflects a mutational change to greater virulence for, in particular, corvids like crows and magpies. That may be true, but Telford Work and Dick Taylor showed years back in Egypt that the WNV strain they were studying was extremely lethal for the local hooded crows. And both the Israeli and the New York viruses have also proved deadly for domestic geese. On the other hand, experimentally infected chickens circulated the virus in blood for up to ten days and developed a few signs of pathology, but none died or showed neurological symptoms.

* * *

As my golfing friend discovered, sentinel chickens continue to do a sterling service, monitoring WNV activity across the USA for the various state health departments and the national Centers for Disease Control and Prevention (CDC). The massive CDC laboratory complex in Atlanta provides technical and field response backup for the US Public Health Service (USPHS) Commissioned Corps, which has regional offices across the nation.

When the personable and flamboyant C Everett Koop was surgeon general of the USA in the Reagan administration, he mandated that USPHS officers wear their elegant white uniform at least once each week. During that time, I had the odd experience of visiting Tom Monath at Fort Collins, Colorado, to find him dressed like an admiral. A prominent arbovirologist, Tom is now in the private sector and has recently been involved in developing a live, attenuated WNV vaccine. The main use of this vaccine so far has been to protect horses. Right at the beginning of the New York outbreak, some racehorse owners moved their prized stock south to Florida. As viruses do, though, WNV soon caught up with them, so vaccination is the better option.

The task of coordinating what is happening with WNV in the USA is the responsibility of the CDC Division of Vector-Borne Diseases. Data is collected weekly from wild birds, sentinel chicken flocks, human cases, veterinary cases and mosquito surveillance. The CDC officers have to be paid, of course, unlike the sentinel chickens, which don't even need to be checked every day. Given the state of contemporary technology and the professionalism of the CDC and associated organisations, this must be one of the best-analysed epidemics in history, challenged only, and for much the same reasons, by what happened in 2009 with the swine influenza pandemic.

Though there is some history of human-to-human spread, through blood transfusion—the US blood supply is checked for WNV—or mother's milk, for example, there is no doubt that WNV is normally spread by the bite of mosquitoes. The principal vectors are various *Culex* species but more than 50 strains of mosquitoes can be infected with this virus in the USA alone. In addition, there is evidence that WNV can be transmitted vertically through the various insect larval stages, and that the virus can overwinter in infected hibernating mosquitoes.

While horses and humans are occasional hosts, it is very clear that WNV will now be maintained permanently in North America by a bird–mosquito life cycle. Right from the outset in July 1999, the virus was detected in mosquitoes, crows and humans. As early as June 1999, the residents of the New York suburb of Queens noticed an unusual number of dead and dying crows. Dead birds were collected over the ensuing months, and of 280 or more that were shown definitely to be WNV positive between August and December, about 90% were American crows. By the end of 2000, evidence of human WNV infection was found in the contiguous states of New Jersey and Connecticut, and it continued to spread north (into Massachusetts and Ontario) and south (into Florida and Louisiana) through the following year. The CDC figures then show a progressive westward distribution, with the first years of high incidence in Colorado and California being 2003 and 2004 respectively. Cases of WNV infection have now been reported from all the states of the Continental USA, north to Alberta and Saskatchewan in Canada, as far east as the Caribbean, and down to Mexico and Argentina, though the disease has not spread to Alaska or Hawaii. Working from the annual CDC figures that have been published since 1999, epidemiologists had (by December 2009) identified neurological symptoms associated with WNV infection in more than 17 000 people, and the virus is considered to have caused at least 1100 deaths.

That rate of geographical spread is completely consistent with the idea of a slow, progressive mosquito-borne transmission from viremic hosts (mainly birds) to previously uninfected avian populations. American crows and raptors (eagles, hawks, owls and vultures) are the likely maintaining vertebrate hosts. While the majority of infected crows are thought to die within three weeks, that still allows plenty of time for mosquitoes to feed, especially if the birds are debilitated by the infection. This is likely to be affecting crow numbers, though it is hard to generate solid figures on overall mortality for any wild bird population. That type of information generally comes from volunteer bird watchers and amateur ornithologists. Such data are, for obvious reasons, much harder to collect and collate in the absence of the databases and resources that are available to salaried public health officers for analysing human populations.

A telling account is, however, available for another corvid, the yellow-billed magpie, a California native that local Audubon Society members recently named as their 2009 bird of the year. According to Holly Ernest, who directs the Wildlife Health and Ecological Genetics Unit in the Veterinary Genetics Laboratory at the University of California, Davis, yellow-billed magpie numbers declined substantially following the arrival of WNV. Field surveys by Ernest and her graduate student Scott Crosbie indicated that there was at least a 20% drop in numbers between 2004 and 2006. The difficulty in coming to a more precise conclusion reflects the fact that earlier estimates of population size are not supported by such solid data. According to the CDC, evidence of extensive, human WNV infection was first detected in California in 2004, when case numbers of meningitis and encephalitis went from two (in 2003) to 289. Over the same interval, 12 000 magpie carcasses were reported to the Californian Department of Health, with almost 80% of those tested being WNV positive.

By 2009, observers from the Audubon Society had the sense that yellow-billed magpies may be coming back, while only 67 reported cases of likely human WNV-associated neurological disease were recorded in California. This could indicate several things. It had been very dry in California, which should mean that there were fewer mosquitoes around. A second possibility is that WNV itself has provided the selective pressure to drive the rapid emergence of a more resistant strain of yellow-billed magpie. If so, given the short time frame, the resistance genes must have already been present in a subset of the population that was first hit by the virus. The third alternative is that the virus itself has mutated to a form that is less lethal for the yellow-billed magpie. Perhaps, as bird numbers fell, it may have been advantageous for the virus to select variants that allowed the maintaining host to be more mobile and thus, perhaps, encounter greater numbers of mosquitoes over a broader geographic range. The tools to differentiate between these different scenarios are there, and I expect that the virologists, on the one hand, and the wildlife geneticists, on the other, are actively pursuing the various possibilities.

6

Ticks, sheep, grouse and the glorious twelfth

ALTHOUGH IT'S DOUBTFUL THAT the birds see it this way, the start of the British grouse-shooting season is famously known as 'the glorious twelfth', since the annual slaughter of the red grouse traditionally begins on 12 August. It's a title that conjures up visions of tweedy aristocratic and upper-class Englishmen, with stout brown boots, deerstalker hats and finely crafted Purdey shotguns, carried 'broken' at the middle to minimise the chance of accidentally shooting someone.

Such scenes were still common in the first decades of the twentieth century but are much rarer now. The world record for 2843 grouse shot at a single site in one day was achieved in August 1913, at Broomhead in Yorkshire, by nine 'guns' (shooters), including Major Arthur Acland-Hood, Sir Ralph Payne-Galloway and Mr RH Rimington-Wilson, all of whom are still recorded in the historic annals of eminent 'sportsmen'. Banging away at birds was clearly heaven for the hyphenated. The world record for killing 575 grouse in a single day goes to Frederick Oliver Robinson, Second Marquess of Ripon. Lord Ripon used three double-barrelled Purdeys and two well-trained loaders. In the off-season, he shot insects to keep his hand and eye in. George Windsor, otherwise know as King George V, blew away 1073 grouse over six days spent at Dawnay Estates in Yorkshire. (More recent English royals seem to favour less lethal outlets, often involving the non-destructive use of horses.)

Another gross bloodletting—the relentless human-hunting season otherwise known as the 'Great' War of 1914–18—obliterated many of the younger members of the traditional grouse-shooting class, along with thousands of their servants. Still, grouse shooting continued at a reasonably high level into the 1920s and 1930s and then went into steady decline. Today, many of the grouse-shooting lodges have disappeared, but the culture still survives, particularly in the Scottish Highlands. According to a 2009 article in the *Scotsman*, this industry provides at least some employment for 6000 people and is worth in the range of £250 million annually to the British economy.

Unlike pheasants, grouse can't be bred under controlled conditions, so the success of any commercial enterprise depends on keeping the wild bird numbers high. As a sport, grouse shooting is very expensive, and the lodges also provide the cheaper alternative of pheasants or clay pigeons. But for a hunter, the grouse are uniquely challenging, both because they are small and because they break from cover flying low and very fast. This is a 'high-end' market, with the likely protagonists now being 'sporting' oil sheiks, industrialists, bankers, financiers and top-echelon criminals rather than land-holding aristocrats. Being a 'gun' for a day on a grouse moor can cost in excess of US$15 000, with some of that going towards luxurious accommodation and dining.

Anything that threatens the grouse supply on the 'glorious twelfth' is a problem for the fragile Highland economy. That's why it became a matter of grave concern when the Highland grouse suddenly started to die in large numbers. The problem was, and is, louping-ill encephalitis, a tick-borne flavivirus that has long been a particular interest of my friend Hugh Reid.

As young veterinary scientists, Hugh and I both wrote University of Edinburgh PhD theses from our collaborative, detailed analysis of louping-ill encephalitis in sheep and lambs. We were working at a place called the Moredun Research Institute, established by Scottish farmers in 1920 to solve disease problems in livestock. Moredun scientists have always been concerned with disease problems that have an impact on the Scottish economy, which explains how Hugh Reid later became involved in events concerning 'the glorious twelfth'.

While there had been a hint early on that louping ill might be responsible for the dying grouse, it was Hugh who went ahead and isolated louping-ill virus from just about every sick or dead grouse on the moor. Veterinary diseases tend to have very descriptive names. Healthy sheep sometimes jump vertically when coming out of an enclosure or through a gate. Sheep can also show this 'louping' (leaping) behaviour just before they go into convulsions and die in the terminal stages of infection with louping-ill virus. Similarly, infected grouse first show evidence of unsteadiness, then paralysis, coma, convulsions and sudden death. That progression from earliest symptoms can take as little as 24–48 hours.

Louping ill is transmitted by the multi-host tick *Ixodes ricinus*, which lives for about three years and can survive perfectly well on the ground. The tick transmits the virus to sheep, and there's also evidence that hares can be involved. But for the virus to be maintained in the long term, tick transmission to sheep and lambs is essential. When this cycle is established on a moor, the result can be the loss of 95% of the grouse population. Grouse become infected when a tick attaches and transmits the virus during the process of taking a blood meal or, when scavenging, the birds eat infected ticks. Though grouse are largely vegetarian, they aren't that discriminating, and insects and bugs do form part of their diet.

Moredun scientists developed a vaccine against louping ill way back in the 1930s, so while there's no easy way to give this product to wild grouse, the birds can be protected by vaccinating sheep and lambs to limit the extent of tick infection. Of course, though the antibodies that are generated in response to the vaccine stop the infection, they have no effect on the ticks and don't prevent them from hitching a ride or from taking a sheep blood meal. This can be exploited by using vaccinated sheep as a 'tick mop': the sheep are put on the pasture, and after enough ticks have come on board, their wool is painted with a topical acaricide (tick-toxic chemical). A final solution is, of course, to remove all the sheep, but that may not be economically feasible, and in any case, it would take about five years before the moor would be considered safe for grouse.

Hugh did some 'virus archaeology' by comparing the gene sequences of louping ill and related virus strains isolated from different regions in the United Kingdom and Europe, and he concluded that the infection probably came to the Scottish Highlands about 400 years ago with the monks who established what became the Scottish sheep industry. Though sheep are less important than they were—with the virtual disappearance of any demand for the coarse 'carpet' wool of the highland breeds—there is still a strong market for Scottish mutton, lamb and, of course, the haggis, which is made from oats, spices, sheep's stomach and other offal. Along with grouse, sheep remain an important part of the Highland economy, though the two species cater to totally different market sectors.

During his investigation of the grouse and louping-ill problem, Hugh Reid spent time talking with those who operate and own the grouse moors, coming to respect them as a unique and well-motivated group of people living close to the land that they and their families have cherished for a very long time. What if altered social values, climate change and/or disease were to bring commercial grouse shooting to an end in the Highlands? Hugh's perception is that a whole traditional way of life would also disappear. But not only that: with the valleys depopulated of people, the moors would return to their former tree-covered state. Lose the open, heather-covered hills, and the Highland landscape is irrevocably changed. That might not be so bad if it were allowed to recover its medieval character, but the certain outcome is that it would become just another dull hill region covered with the usual monoculture of pine plantations.

There's an odd paradox when it comes to protecting wildlife, in that the numbers can sometimes be best preserved by sequestering rural habitats from 'development' and allowing controlled hunting. Providing an economic motive for sustainability means that there are people who can both be advocates and provide the resources to undertake the necessary monitoring and protective measures. That's the premise for the Ducks Unlimited organisation in the USA, and particularly with free-flying birds that do not restrict their territory to game reserves, it is clearly one way to go. Prohibiting the hunting of particular species is a possibility, but it also disturbs the natural balance if one species can be hunted and another cannot. Sometimes, like the famous American experiment with banning the sale of alcohol, it just makes the problem worse and turns ordinary people into petty criminals.

The protected hen harrier, for example, eats grouse chicks, so the fact that hen harrier numbers do not seem to be recovering from threatened levels inclines concerned birders to the view that these hawks are being killed illegally. Perhaps, but gamekeepers on the grouse moors also deserve credit for promoting the survival of wild birds as they work to control predators, particularly foxes, stoat and carrion crows. One solution that's been suggested is to breed hen harriers in captivity, then release them after the grouse chicks have grown to a size where they are no longer prey. Another alternative is to provide diversionary feeding. In effect, what's being said here is that we must now manage these two wildlife populations so that an appropriate natural balance can be maintained. That's the challenge we're facing globally, in fact, as more and more species become threatened by rapid climate change and the ever-increasing intrusion of human beings, along with the toxins, debris and general degradation of habitat that are the inevitable consequences of our carelessness and presence in large numbers.

Whether we like the idea or not, humanity has the job of managing the planet, a process that can be in direct conflict with long-held beliefs and well-established practices, on the one hand, and with the urban romanticisation and anthropomorphisation of animals (the Bambi syndrome), on the other. We need to change how we do things, but in ways that are driven as much by a respect for science and the realities of nature as by emotion. That's one reason why, when it comes to birds or anything else, the more we understand the underlying data, and the more we design econom-ically realistic solutions, the better the outcomes are likely to be. Sophisticated advocacy organisations know that, of course, and lobby politicians accordingly.

7

Flu flies

I had a little bird and its name was Enza,
I opened the window and in flew Enza.

MANY OF US WILL have heard this rhyme, originally sung by children during the terrible 1918–19 'Spanish flu' pandemic. People were dying all around them. About one in 40 of those who contracted this infection did not survive, with kids and young parents being among the victims. The Spanish influenza is still the worst episode of a fast-spreading, lethal plague in modern times. And while most long-distance travel in those distant years was on slow ships, flu does fly now, even without birds.

The ditty about Enza was strangely prophetic: way back then, we didn't know that there is a vast reservoir of influenza virus infection in birds. While a very limited number of influenza variants circulate within human populations, it's waterfowl, not humans, that maintain an extraordinary spectrum of these viruses in nature. In fact, if we look at the interface between birds and humans purely from the aspect of infectious disease, influenza is by far the most important subject that we have to discuss. As such, it merits more than one chapter in this book. And it needs to be considered from both sides of the complex avian–human relationship.

Most of what follows in this chapter is about the influenza viruses and influenza in general, and the scientific evidence presented is drawn more from our understanding of what happens in placental mammals than in birds. It's necessary to know something of the virus, its genetics and how the vertebrate immune response deals with this infection. Research on disease processes in humans and their experimental surrogates—particularly mice and (with influenza) ferrets—commands vastly greater financial resources than are available for any study of wild or domestic birds. As a consequence, much of our thinking about the actual disease process in avian influenza is based in evidence gleaned from studying the problem in people and various furry and hairy (rather than feathered) laboratory animal species. Like humans, birds clearly suffer influenza infections of varying severity, though they can't tell us directly how they feel.

Everyone has had the flu, and most of us don't confuse influenza with 'just a bad cold'. There are several reasons for this. 'Seasonal' influenza hits regularly in extensive, rapidly spreading global epidemics that get plenty of media coverage, since some people end up in critical care beds (or worse). A vague sense of guilt and trepidation can arise on these occasions as we recall that we were just too busy and didn't get around to taking the current flu vaccine. Birds don't have that issue to deal with, of course, but it may be a

consideration for poultry producers who have neglected to vaccinate their flocks in the face of a potential outbreak.

The main reason we don't confuse flu with a bad cold, though,
is that it's just a whole lot nastier than the 'usual suspects' (para-
myxoviruses, coronaviruses, rhinoviruses) that cause coughs and
sneezes. The first symptoms of fever, headache, general dullness
and so forth are pretty typical for a lot of bad virus infections,
including West Nile, Japanese encephalitis and yellow fever. Even
before that initial flu phase hits and we're infected but still feel
reasonably OK, we can be spreading the virus around to friends,
family and anyone who happens to be near us on a plane, train
or bus. Data from the CDC, which tracks 'seasonal' epidemics in
the USA, show the virus travelling to every state in the space of
4–6 weeks. Contrast that with the four years it took for West Nile
virus to get from New York to California using a bird–mosquito
infection cycle (see Chapter 5).

Human flu flies as we fly, with the speed of jet travel. And, if
we're incubating the virus when we start our trip, we soon know
why we have been feeling a bit feverish and dull. The more severe
symptoms start with coughing and spluttering, then muscles ache,
our breath gets raspy, and we may find ourselves struggling to get
enough air in through our 'consolidated' lungs. The combination
of virus damage and inflammation can make the delicate, pink
lung tissue look like a piece of liver. That's one of the effects that
can put us into an intensive care bed. Even after the respiratory
phase resolves, we may feel depressed and down for weeks.

This is not a good disease, and it's definitely to be avoided.
Compromise the cells deep in the lung that are involved in O_2/
CO_2 exchange and any air-breathing vertebrate is in big trouble.
During the recent H1N1 'swine' flu outbreak, intensive care
physicians used heart-lung machines to supply oxygenated blood
and get severely affected patients through the crisis. The good
news is that some, at least, recovered completely. Birds in the
wild have, of course, no possibility of surviving such a severe

respiratory crisis. Even a relatively mild infection would severely limit their capacity to fly and (with many species) to feed.

At the extreme end of the spectrum, there's no doubt about the part that influenza virus played in causing some of the recent hyper-acute 'bird flu' deaths in humans that University of Hong Kong physician and researcher Malik Peiris has attributed to 'cytokine shock'. It's still early days in our understanding of this 'syndrome', to use the medical jargon. What seems to be happening is a massive overreaction of the relatively non-specific host-response elements that kick in as soon as the respiratory tract epithelial layer becomes infected by an incoming flu virus.

This 'immediate' response comes first from the infected cells themselves as they produce various defence molecules like the inter-ferons. Then a specialised army of bloodborne 'warriors' mobilises rapidly to the site of infection. These 'inflammatory' white blood cells—the neutrophils, macrophages, dendritic cells and 'natural killers'—are major players in what we call the 'innate immune system'. Among other functions, they spew out a lot of potentially toxic molecules (called cytokines and chemokines) into the sur-rounding tissue, with any excess being carried away in the blood. All that is completely normal and helps to keep the infection in check until the more specific (or 'adaptive') immune response, the part we 'prime' by vaccination, can take over and eliminate the virus, usually after about 7–10 days.

Bloodborne cytokines also promote fever and lethargy, both of which are protective measures that induce us to slow down. In addition, these 'chemical mediators' are the likely cause of any muscle pain and later depression. It's always possible to have too much of a good thing. Cytokine shock results when the generally beneficial cytokine/chemokine response is just too rapid and too effulgent. The toxic molecules that damage the invader can also compromise the integrity of the blood vessel walls, causing vas-cular breakdown and massive fluid leakage into infected tissues. Patients can drown in their own lung fluid. Just how big a part such effects play in the hyper-acute deaths seen in some poultry outbreaks is simply unknown

What happens when a particular strain of influenza virus has the capacity to cause this type of extravagant response is that we see acute deaths in the otherwise healthy young. There's been a little of that with the generally mild 'swine' influenza that crossed over into human populations in 2009, with a number of particularly severe (and poorly understood) cases in pregnant women. Similar atypical 'shock' symptoms with an unusual age–susceptibility profile were prominent during the 1918–19 'Spanish' flu pandemic that killed 40–100 million people globally. The wide variation in estimated mortality reflects the state of global communications back then, particularly when it came to conducting meaningful censuses of isolated villages in far-flung colonies.

The realisation that the influenza A viruses we've been discussing up till now are maintained in nature by birds, particularly aquatic birds, did not come until many years later, when the first influenza virus was isolated (from pigs) by Richard E Shope. Back in 1931, when Dick Shope was doing his stuff at the Rockefeller Institute branch laboratory in Princeton, New Jersey, the primary evidence that an infection was caused by a virus was that the organism could not be grown (or cultured) like a bacterium or fungus in some sort of cell-free 'broth' or on, say, blood agar plates, and that it was very small.

What Dick Shope did was to grind up some tissue from an infected pig in order to make a filtrate, which he then used to transfer a relatively mild form of pneumonia to other pigs. Not long after, in 1933, a group working at the Medical Research Council laboratories in Mill Hill, London, made the first human isolate by dropping infected material down the noses of ferrets. The Melbourne-based chick embryo pioneer Macfarlane Burnet told the story that he was visiting Mill Hill when he encountered a guy rushing down the corridor shouting, 'The ferrets sneezed, the ferrets sneezed!' Now there's the excitement of science for you! I tell young scientists that they should hope that the 'ferrets sneeze' at least once for them during their career. Like Burnet, the human members of the Mill Hill influenza team were all honoured with knighthoods by the British monarch, though no monument has been erected to recognise the contribution made by the humble ferret.

More than a decade elapsed between the end of the catastrophic 1918–19 pandemic and the isolation of the first influenza viruses by Richard Shope, and then by the British team of Wilson Smith, Christopher Andrewes, Patrick Laidlaw and Charles Stuart-Harris. After that, the detailed characterisation of the virus took a further 30–50 years, and we are still discovering aspects of what the eight influenza virus genes actually do. With modern technology, though, the cause of the terrifying 2002 SARS outbreak, which killed about 800 people, was definitively identified in a matter of weeks. Like Shope's 1931 influenza isolate, this was a pathogen (a bat coronavirus) that we'd never seen before.

* * *

Before we can take this discussion of influenza further, we need a bit of technical background on the influenza viruses. The influenza spectrum is divided into A, B and C viruses, all members of the myxovirus family. Only the A and B viruses circulate in humans, while it is the A viruses that naturally infect aquatic birds.

The genetic information for the influenza A viruses is carried in their nucleic acids (RNA). Because of the lack of an adequate 'proofreading' system for quality control, this RNA undergoes mutational change at a very high rate. Such spontaneous mutants can be selected by either the antibodies or (less commonly) the 'killer' T cells, which, after about 7–10 days, have expanded in numbers and quality to the extent that they control the original infection. Working together, these two different 'adaptive' immune mechanisms tend to ensure that an emerging virus variant will not persist within the individual. However, any mutant that does

manage to 'sneak past' these defence mechanisms may now—
if it has a high degree of 'fitness'—infect susceptible contacts. The
occasional 'success' (for the virus) of this process gives rise to
the novel, 'seasonal' influenza viruses that cause global epidemics
at one- or two-year intervals. Influenza virologists and epidemiologists refer to such changes as 'antigenic drift', reflecting the
fact that the 'antigens' (proteins) targeted for immune attack are
so modified that they now seem novel. Rather than being immediately recognised by pre-existing antibodies or triggering a
'recall' of memory B cells and T cells that have been induced by
prior infection or vaccination, the mutated (or 'drifted') proteins
expressed by novel virus strains can only be eliminated by a new
and much slower 'primary' response.

In addition, the genes of the influenza A viruses are organised
in eight different segments. As a consequence, if one cell is infected
concurrently with two different influenza A strains, the new
progeny viruses that result can include 'reassortants' that have
some genes from one virus and some from the other. This mixing
experiment is easily done in the laboratory and was discovered in
1951 by Margaret Edney (later Sabine) while working as a very
junior scientist in the laboratory of FM Burnet. Such 'antigenic
shift' can result in the emergence of novel pandemic strains.

While working with Burnet, Edney would also have learnt the
haemagglutination inhibition (HI) technique, a method that was
first developed for influenza (in 1941, by George K Hirst at the
Rockefeller Institute) and was soon well established in research
and diagnostic laboratories. The HI method detects antibodies
that prevent influenza viruses from binding to red blood cells
(RBCs), such binding being a defining characteristic of the myxovirus family, which includes influenza. Most flu labs have a goose,
a duck or a rooster that they bleed regularly for this purpose. A
blood sample is taken from the wing vein of a normal bird, and
the RBCs are separated and suspended in salt solution. Shake the
mix in a test tube, then set it on a lab bench or put small volumes

into the 96 wells of a culture plate, and the RBCs soon drop to the bottom to make a tight pellet. Mix in any flu strain at the stage when the RBCs are in suspension and the haemagglutinin (H) protein on the virus surface cross-links the RBCs (haemagglutinates) so that instead of settling as a pellet, they sink to form a fuzzy lattice. Adding immune serum containing antibodies that bind specifically to the influenza H molecule blocks the interaction and stops lattice formation, thus the term haemagglutination inhibition (HI). The difference between the lattice and the pellet is easily read by eye, so it's a simple test to do and it is also very sensitive.

Because such antibodies are incredibly specific for a particular site on the virus H protein, knowing their concentration can tell us both the identity of the infecting strain and also give some idea of how long ago the individual was exposed. Furthermore, a serum HI antibody titre of greater than 40—the dilution below which the lattice–pellet transition occurs—is associated with at least a 50% reduction in the risk of infection. Measuring the HI level allows us to check the efficacy of any influenza vaccine we've given to people, chickens, pigs or horses. Antibody tests also provide access to a lifelong history of natural exposure in, for example, sentinel chickens and elderly humans. Each of us has a unique story to tell, and part of our story can be read by measuring serum antibody titres.

In nature, aquatic birds are the maintaining hosts for a very broad range of influenza viruses expressing, in combination, some 16 different H and 9 neuraminidase (N) molecules. People suffer occasional H5N1, H7N7 and H9N2 infections, but at least over the past century or so, only the H1N1, H2N2 and H3N2 viruses have become established in human populations. Influenza viruses 'jump' from birds into other species, including leopards, seals, whales and domestic cats. Variants of both the H7N7 and H3N8 viruses are maintained in horses, and horses are considered to be the source of a further H3N8 transition into racing greyhounds. Since this happened (in 1999), the H3N8 virus has become established in dogs of all shapes and sizes and is clearly being maintained by dog-to-dog spread.

Back in 1959, an H7N7 virus that probably came from infected gulls killed more than 400 harbour seals in the Boston area. The latest fatal H7N7 human case was in the Netherlands in February 2003, when some 289 people were infected, largely during the course of efforts to control an outbreak affecting commercial chicken populations. The exact origin was not identified, but the crossover is thought to have occurred when free-range chicken and turkey flocks were infected by migrating waterfowl.

Though the influenza A viruses naturally infect waterbirds, that doesn't mean we normally catch flu from embracing a duck, a goose or a swan. People are much more dangerous. But as in the case of the Netherlands H7N7 outbreak, we can become infected as a consequence of being around large numbers of diseased chickens. That was certainly the case for the veterinarian who died and for a number of other people who developed an unusual influenza virus–induced conjunctivitis (inflammation of the outer eye), but this particular virus never changed in a way that allowed it to spread in human populations. Where, then, did the H1N1, H2N2 and H3N2 viruses that became established in people originate?

Both the H2N2 'Asian' (1957) and the H3N2 'Hong Kong' (1968) pandemic flu viruses are considered to have arisen by reassortment, or 'antigenic shift', between a human virus and a strain circulating in ducks. The 'parents' that provided the genes for the H3N2 virus are thought to be the human H2N2 and a duck H3N8. When it first struck in 1957, the H2N2 virus initiated a global pandemic that killed an estimated one million people. It was then 'displaced' by the H3N2 'Hong Kong' virus and is no longer circulating in human populations, though there are concerns that it is still lurking out there in other species and may come back. Since 1968, 'drifted' H3N2 and H1N1 viruses have been responsible for sequential, 'seasonal' epidemics that regularly cause more

than 30 000 deaths in the USA alone. The frail elderly are generally the most vulnerable, and as pneumonia provides a relatively
benign exit, influenza used to be called 'the old man's friend'.

And there's the terrible H1N1 'Spanish' flu, the virus that
may have killed as many as 100 million people in 1918–19, when
the global population was only a third the size of that today. Not
having the virus variant that caused the problem at the time, the
only evidence that we had for the identity of this devastating
pathogen came much later from analysing the 'footprints' of prior
infection, the circulating antibodies that could be measured in the
serum of long-term survivors. That allowed classification of the
1918 virus as an H1N1 virus with some similarities to strains that
have continued to circulate in pigs, but we still had no evidence
of its exact identity. Then, in 1977, there was a fresh H1N1 outbreak in humans, possibly caused by an escaped laboratory strain.
Remarkably, elderly people who had survived the 1918 pandemic
were relatively protected. Those antibody molecules that defeat
influenza viruses at the point of entry were still around, showing
us that immunological memory can last for more than five decades.

The solution to the identity of the 1918 virus came with the
advent of the polymerase chain reaction (PCR) technique, which
allows the expansion of very small amounts of RNA or DNA to
give sufficient material for obtaining a gene sequence. Developed
by Californian scientist Kary B Mullis, this is the same method
that's used to identify rapists and establish paternity.

Using PCR, Jeffrey Taubenberger at the US Armed Forces
Institute of Pathology (AFIP) in Washington, DC, set out to
discover the genetic sequence of the 1918 virus in the hope of
determining what made it so lethal. Starting with formalin-fixed
lung tissue (an AFIP museum specimen) from a young recruit who
died very acutely in 1918, Jeff was able to recover a good measure
of the virus genome. The sequences were essentially completed
when Johan Hultin provided further material that he'd isolated
from the lungs of a young woman who had again died relatively

early in the course of the infection and was buried quickly in the Alaskan permafrost. Somewhat on the heavy side, both her body fat and the fact of being immediately frozen had protected the viral RNA to the extent that the PCR reaction could proceed with sufficient fidelity.

From 1997, the research papers that identified the 'lost' 1918 virus started to appear in the leading scientific journals. Looking at the gene sequences, it soon became very obvious that the 1918 pandemic was caused by a mutated bird virus. Not only that: when this terrible pathogen was reconstructed and given to monkeys housed under very high security conditions, it caused extremely severe disease and stimulated the 'cytokine storm' that Malik Peiris had identified as the possible cause of rapid mortality.

What is it that changes when an avian influenza virus (like the 1918 H1N1) suddenly starts to spread between humans? As the experience with the Netherlands H7N7 outbreak showed, it isn't enough for the virus to simply jump from one species to another. Over the years, much of the focus on explaining the bird to human 'adaption' for the influenza A viruses has focused on the interaction of the virus H molecule and the sialic acids (sugars) found on the surface of respiratory tract cells. While the basic characteristics of cells have very many common characteristics (whether they be worm, mosquito, bird or human cells), there are also features that are unique to a particular species, genus, family or phylum. The sialic acids found on the surface of human and avian respiratory tract cells are variations on a theme, but they're not identical. When growing influenza viruses in tissue culture under the right conditions, though, it's pretty easy to get them to switch from one form to the other, and it's a bit surprising that this crossover doesn't occur more often.

The long-term evidence suggests that most bird-to-human influenza 'jumps' occur somewhere in South-East Asia. The reason is that these regions are consistently warm and damp, and people still live more traditional lives, where they are in close contact with their animals. There are also very large numbers of aquatic birds, both wild and domestic. Partly because the very first influenza virus was isolated from pigs, there's been a lot of emphasis on the idea that pigs act as the 'mixing vessel' for influenza virus reassortment, the 'antigenic shift' discussed earlier in this chapter. Pigs are thought to have been key intermediates for transfer of both the H2N2 and H3N2 pandemic strains, though because we have none of the 'precursor' viruses for the 1918 'Spanish' H1N1, it's not possible to say exactly what occurred back then.

Why should pigs be important? Dick Shope thought there might be a role for 'carrier' lungworms, which infect both pigs and humans, but that idea has now been discarded. The basic reason is that pig lung cells express both the 'avian' and 'human' forms of sialic acid, allowing the double infection with a bird and a human virus that's essential before any 'reassortment' can occur. The idea that influenza viruses jump from pigs to humans has become familiar to all of us since the H1N1 'swine' flu pandemic that kicked off in Mexico at the beginning of 2009. A virus that had been circulating for years in pigs suddenly changed in a way that made it more infectious for people.

When it comes to crossing species barriers, it's clearly the case that flu can swing both ways. Some components of the 2009 'swine' HIN1 that caused the first human influenza pandemic of the twenty-first century are derived from the H1N1 'Spanish' flu that went into humans and pigs about the same time in 1917–18. We don't know whether virus from the pigs infected people way back then or if it was the other way round, and it's likely that this will remain a mystery. After the catastrophe of 1918–19, milder, seasonal H1N1 epidemics continued up until the mid-1920s or so, and the virus eventually disappeared from human populations. It seems, however, that pigs continued to maintain that H1N1 virus. What the virologists and physicians worried about with the recent H1N1 'swine' flu was that as it circulated in people, it might mutate to greater virulence. There's nothing written in stone that tells us how to predict novel influenza pandemics. Though we can be pretty sure that influenza A viruses will occasionally jump into human populations, we don't know which particular strain will make the transition, when this might occur, or how severe and extensive the disease might be in us. Chance rules, and as with the risk of war, we need to be both vigilant and to keep our defences in good shape. That might sound a bit over the top, but bear in mind that the 1918–19 Spanish flu claimed many more lives than were destroyed by the collective insanity of World War I. Our general level of awareness was raised for a time by the 60% human mortality rate associated with the H5N1 bird flu epizootic (an epidemic in animals rather than people), but we tend to forget about such possibilities when they're no longer in the news.

In December 2011, though, that changed, with extensive reporting on the controversy associated with the publication of genetic changes in H5N1 viruses that are lethal for chickens and, after being passaged serially in ferrets, acquire the capacity to transmit naturally in that species. When it comes to spreading flu, ferrets and humans are very similar. Combine that with the fact that the relatively mild 2009 'swine' H1N1 went round the world in less than five months, and it's obvious that we cannot afford to be relaxed and comfortable when it comes to influenza.

8

Bird flu: from Hong Kong to Qinghai Lake and beyond

HAVING DISCUSSED INFLUENZA IN general, it's time to focus on the birds. The first thing we have to realise is that, being natural and generally mild infections of a whole range of waterfowl, there is no way that these potentially lethal pathogens can ever be eliminated unless our small planet turns into a dry, lifeless rock. So long as we share the earth and oceans with our feathered cousins, mammals such as ourselves (and seals, whales, pigs, horses, leopards, racing greyhounds and so on) have to live with the certainty that novel influenza A viruses will occasionally 'jump' from wildlife reservoirs, sometimes with disastrous consequences.

Aquatic birds play a key role because, unlike many viruses, influenza survives happily in fresh water, which means that ponds, lakes and dams are a major source of cross-infection. Ducks, geese, flamingos, cranes, waterhen and so on all have to drink. Even passerines that aren't normally swimmers and divers are at risk and can, for instance, become infected and spread the virus to poorly protected chicken houses. Cross-infection becomes more likely under dry or drought conditions, when sparrows, swallows, starlings and waterfowl come close together at reduced bodies of water.

Avian diversity is an essential part of the mix. The same virus that kills one bird species very quickly may cause a subclinical, long-term, persistent infection in another. In general, it's a very bad idea for commercial chicken producers to keep a pond where they raise a few domestic geese or ducks. Visits from their free-flying, migrating relatives can introduce an unwelcome influenza guest that first infects the local water birds, then spreads to cause chicken mayhem.

Maintaining very large numbers of free-range chickens and turkeys is also risky, as the probability of exposure to wild birds is much greater under field conditions. That's thought by many to describe what happened with a 2003 H7N7 outbreak in the Netherlands. Public concern for the welfare of chickens that are caged for life had led to a 'deinstitutionalisation' of the local poultry industry, the emphasis being on allowing the birds a more natural lifestyle. The disease spread to poultry in Belgium and Germany, was responsible for one human death and led to the culling of some 33 million domestic birds.

Looking, say, at wild-caught Canada geese before the fall migration, as many as 30% of juveniles can be shedding one or other strain of influenza virus. You've never seen, or heard, of a goose or a duck with what looks like the flu or even a bit of a cold? The reason is that influenza is generally a relatively mild infection of the avian gastrointestinal tract (rather than respiratory tract). While the goose or duck immune system works perfectly well and eventually gets rid of the pathogen, the virus can be detected in gut contents for 5–10 days without causing any obvious problem for the birds. And it's not only wild birds that are involved: influenza virus can, for instance, be excreted from the common cloacal opening of healthy domestic ducks for as long as 17 days after experimental infection. The pasty, white mixture of faeces and urinary tract excretions that we've all experienced sometime or other as a 'gift' from on high is highly infectious when voided into water, and survives well in wet bird droppings on the ground.

The duration of virus excretion, together with both the variety and the enormous numbers of wild birds that are involved, means that there is little cause for the viruses to change in order to maintain their avian transmission cycle. That's very different from the situation with long-lived humans, where durable antibody protection in those who have been vaccinated or recovered from influenza exacerbates the problem of limited numbers in the potentially susceptible pool. The virus will just die out if it can't keep transmitting from person to person. The net consequence is that influenza viruses that are being maintained in human populations are under constant selective pressure to alter their outer-coat H and N proteins and escape from antibody-mediated immune control. That 'antigenic drift' effect is normally minimal for influenza infections of wild birds, and the viruses can remain essentially unchanged for decades.

The same can also be true for domestic birds: relatively innocuous, 'low-pathogenicity' (low-path) influenza A viruses can circulate in chicken populations for years without causing much of a problem. Producers don't generally bother about vaccination, but these viruses sometimes mutate to a 'high-pathogenicity' (high-path) form and cause massive, lethal outbreaks. High-path influenza in birds is completely different from the normal situation for ducks, geese and chickens infected with the low-path strains. These high-path viruses are systemic, meaning that they spread in the blood and grow in all organs, including the brain and the lungs. Large numbers become very debilitated, show haemorrhages in the skin and legs, turn blue (cyanotic due to lack of oxygen) and die very quickly. The disease can be extraordinarily severe and is quite terrifying to watch. The very natural response is to ask: what if that were us? Those who deal with such avian outbreaks have been the first to warn about the potential danger that influenza poses to crowded, and highly mobile, human populations.

According to the World Health Organization (WHO), there were 21 avian high-path flu incidents reported between 1959 and 2003, occurring in Europe, Asia, Australia, and North and South America. This low- to high-path 'switch' has been seen only for H5 and H7 viruses, though low-path H9N2 and H6N2 viruses can sometimes cause problems as co-infections with something else. The WHO figure for 'high-path' transitions is undoubtedly an underestimate, as the likelihood that such an event will be recognised depends on the sophistication of both farmers and the local veterinary services. Five of the 21 were in Australia (caused by H7N7, H7N3 and H7N4 viruses), where general awareness among producers and regulatory authorities that exotic infections can cause enormous economic loss means that there are well-established state and national animal virus diagnostic laboratories.

For the virologists and other flu specialists, perhaps the most dramatic wake-up call prior to the H5N1 bird flu story came with the 1983 outbreak caused by a high-path H5N2 virus in Pennsylvania chicken flocks. Suddenly, seemingly out of nowhere, large numbers of chickens started to die in commercial production facilities. The disease spread to the neighbouring states of New Jersey and Maryland, and was ultimately handled by culling more than 17 million birds at a direct cost of some US$60 million. Humans did not become infected, and though the original virus may have come from wild birds, live bird markets were thought to be the likely focus of dissemination throughout the chicken industry. Ducks, geese and so forth were infected experimentally under laboratory conditions, but none of those species either developed severe symptoms or excreted unusually large amounts of virus. Then the high-path chicken experience was repeated in Mexico in 1993, and in 2006, 60 ostriches died from an H5N2 infection on a farm in South Africa. Thinking about this, it becomes very obvious that the process of genetic change that causes any influenza A virus to switch from being mild to severe can occur any place at any time.

What frightened the virologists was that the transition from low-path to high-path resulted from a single point mutation in the viral RNA. That is, the 1983 H5N2 virus became extraordinarily lethal for chickens as a consequence of just one amino acid change in the virus H protein. This was long before Jeff Taubenberger and Johan Hultin reconstructed the 1918 pandemic virus, and it was at the forefront of everyone's mind that such a minimal modification could have occurred back then to cause the virus to 'jump' from some other species (pigs or birds perhaps) into humans. The serial-passage H5N1 ferret study mentioned in chapter 7 indicated that five mutational changes may be sufficient to allow this avian virus, which is highly lethal for the relatively few people who have been infected, to transmit readily between humans. Then, though, there's also the possibility of reassortment to allow the emergence of a virus that has some genes from a strain that infects birds, and others from an isolate that transmits readily between people. That clearly happened at least twice during the twentieth century. When the Dutch authorities had to deal with the 2003 H7N7 outbreak in chickens and turkeys that also caused some human infections, the first thing they did was to make sure that those 'in contact' were given the standard 'seasonal' influenza vaccine, to minimise the risk of concurrent infections with a human and a bird strain, the necessary prerequisite for gene reassortment. Some of those handling the infected birds were initially reluctant to take the influenza antiviral drug Tamiflu, but that changed after one of the veterinarians died from the infection.

The realisation that a lethal flu outbreak in domestic poultry can constitute a direct threat to human wellbeing first entered the broader consciousness when, between May and December 1997, a high-path H5N1 virus caused six deaths and otherwise severe disease in a total of 18 Hong Kong residents. This was the first time that an H5N1 strain had been shown to make the jump and cause major clinical problems in humans. All the evidence indicated that the virus, which had been causing deaths in domestic chickens from March, was in some way contracted from infected birds and did not spread between people. Comprehensive testing of birds in the wild, on farms, from zoological gardens and in the live bird markets that bring city people into direct contact with ducks, geese, chickens and so forth showed evidence of widespread involvement. A report by Ken Shortridge, who was then the senior resident influenza researcher in Hong Kong, relates that the H5N1 virus was isolated from 2.4% of ducks, 2.5% of geese and 21% of domestic chickens.

The Chinese tradition is to eat meat that is as fresh as possible. As a consequence, Shortridge reckoned that in Hong Kong in 1997, there were about 1000 urban live-bird markets. Apart from the obvious dangers associated with killing and dressing an infected bird for the table, the possibility of infection from contaminated bird droppings and local water sources was also suggested by the fact that some of the cases clustered in neighbourhoods near the bird markets. From late December, the epidemic was brought to an end with the culling of some 1.5 million domestic poultry.

During 1998, I visited my US-based colleague, the avian influenza expert Rob Webster, who was then spending a few months every year helping with the 'influenza virus watch' program operated by Shortridge and his colleagues at the University of Hong Kong. Rob took me along to one of the live-bird markets. By that time, ducks and geese had been banned, but there were many cages of chickens and quite a few 'loose' birds roaming free. Then there were cages of tiny quail, which were often placed right next to, or even directly under, the chickens. Looking closely, all the quail had ruffled feathers, and even for a lapsed veterinarian like me, it wasn't hard to see that they were not in optimal health. Rob assured me that he could pick any one of those fluffed-up little birds, take a swab from the cloaca and isolate one or other influenza virus.

It was a lesson in epidemiology or, to be more correct, epizootiology. The chickens and quail sell fast, but they in turn infect pheasant, chukar and guineafowl that move more slowly (in the commercial sense) and hang around to infect the next batch of chickens freshly arrived from the farm. At that stage, there was no H5N1 flu anywhere in Hong Kong, at least so far as anyone was aware, but H5N1 was never the whole game for the local avian world. The widespread testing during the 1997 outbreak also turned up several H9 strains that were circulating at a lower level (0.9% in ducks, 0.6% in geese, 4.1% in chickens and 3 isolates from pigeons). One sampling of a live-bird market showed that 36% of the birds were H9 positive! The same situation no doubt applies in many situations where large numbers of birds are brought together at close quarters, but the H9 influenza strains were low-path viruses that were not causing problems, so nobody was too bothered.

The kill-off in Hong Kong was not, sad to say, the end of the story for the high-path H5N1 viruses. An April 2010 news story, for instance, reported that a 27-year-old man had died in Cambodia from the H5N1 bird flu virus. He was the tenth person in the country to be infected with the virus, and the eighth person to die from it. The report went on to caution people to be on the lookout for sick poultry and to report such incidents to the authorities.

As of March 2010, the WHO had recorded 486 human cases of H5N1 bird flu, with 287 deaths, a mortality rate of around 60%. By October, the figures were more like 500 and 300. Apart from Cambodia and the 1997 outbreak in Hong Kong, people have died on the Chinese mainland and in Azerbaijan, Egypt, Indonesia, Laos, Indonesia, Nigeria, Pakistan, Thailand, Turkey and Vietnam.

Clearly this virus, though highly lethal, is not very infectious for humans and has yet to make that 'jump' that allows it to spread between people. But, despite the low infectivity, any virus that causes 60% mortality raises obvious concern. That's especially true for influenza, which has a well-known proclivity for mutation (or gene reassortment) to extend its host range.

The current high-path H5N1 virus does, for example, cross readily from birds into cats. Domestic cats show a high incidence of infection in areas where the H5N1 virus is killing chickens, and both tigers and leopards have died in zoos, perhaps as a result of being fed infected chicken carcasses. There is as yet no evidence that cats spread the disease to humans, but there is concern that they could act as a 'mixing vessel' (as was long suspected for pigs) to allow strains that are circulating concurrently in birds and people to meet up in the same infected lung cell.

When it comes to human infections, what is thought to have happened with the relatively few unfortunates who have developed H5N1-induced disease is that they received a very large dose directly from infected birds. It now seems that this form of influenza only happens when aerosolised virus penetrates to the farthest ends of the human lung, the small bronchi and alveoli. The team led by virologist Yoshi Kawaoka (who has labs at both the University of Madison, Wisconsin, and the University of Tokyo) has shown that while cells with the 'human' form of sialic acid dominate our upper respiratory tract, some of our deep lung epithelium also expresses the 'avian' form, the preferred receptor for the H5N1 virus.

A classic scenario involves a small farmer in an Asian village who knows that his chicken and duck flocks will be killed and then burnt or buried by regulatory authorities if they show any signs of high-path H5N1 bird flu. At the first hint of infection, he decides to distribute the birds around his extended family so that everyone can enjoy at least one good dinner. His young son stuffs a couple of the better-looking chickens down the front of his shirt, and hops on his bike to take them to Auntie's place. The infected birds breathe almost directly into his face, exposing him to a massive dose of high-path H5N1 influenza virus. Pedalling vigorously, he draws the virus-tainted air into the depths of his lungs. The appalling result is that the child dies of pneumonia 5–10 days later.

Apart from the death of the boy, there are many elements of personal tragedy in this little tale. Among them is that killing large numbers of infected and in-contact birds reduces the supply of high-value protein to people who live on a nutritional knife-edge. Currently, in a world where Westerners are dying prematurely of obesity, there are thought to be just under a billion people who simply do not get enough to eat each day. At least 500 million chickens have been destroyed during the continuing high-path H5N1 epizootic, either by the infection itself or as a result of culling to control virus spread.

Then there's the loss of income as bird populations are slaughtered, reducing, among other things, the financial flexibility that frees men and women from unrelenting drudgery and allows children to be educated. In poorer countries, people's fates and lifestyles are still intimately linked to the health of their animals. According to an old African saying, 'If the cattle die, the people die also.' That may be just as true for chickens, particularly in Asia. As flu spreads, so does a great deal of human suffering, and it is not necessarily associated with people contracting the disease. As we wait for the virus to make the 'jump' that will allow global spread, those of us in the West keep watch, but only for a time; once the journalists and editors tire of the story and switch to something else, we soon lose interest. Media images of jumpsuit-clad, masked men slaughtering chickens and burying them in pits can't rival the visual drama of flood, fire, earthquake or tsunami.

The absence of TV and newspaper coverage does not, though, mean that the problem has gone away. Fortunately, we have organisations like the WHO, the CDC, the US National Institutes of

Health, and the US Department of Agriculture, which continue to watch the situation very closely. Those who attack organisations like the United Nations (UN) and the idea of global cooperation think in terms of parochial politics, not international science. By choice or from ignorance, they are clearly unaware of the massive risk posed by diseases like influenza and the good job that UN agencies, like the WHO, do in looking out for our interests.

In general, both our capacity to monitor infectious disease outbreaks and our understanding of the basic science have been evolving with incredible speed. Rapid technological advances mean, for example, that genomes can be expanded by PCR and sequenced in a day or two to identify an influenza A virus that is causing, say, an outbreak in poultry or a severe human case. Much of that work is done within the six WHO Collaborating Centres, located in Britain, China, Japan, Australia and the USA. These in turn link back to a much larger network of government- and university-based laboratories. Our capacity to follow what is happening globally in 'real time' is probably more sophisticated for influenza than for any other infection, with the possible exception of HIV/AIDS. As things stand, though, influenza is potentially much more dangerous, because of its proclivity for extremely rapid respiratory spread.

* * *

The origin of the 1997 high-path H5N1 Hong Kong outbreak was ultimately traced back to a 1996 virus isolated from domestic geese in Guangdong, China. Since then, despite massive efforts at control, including some ill-advised bird-vaccination programs, lethal H5N1 strains (or 'clades') have continued to spread and to evolve genetically. Hui-Ling Yen, Guan Li, Malik Peiris and Rob Webster summarise the situation for commercially raised birds as follows:

Previously reported high-path outbreaks of H5 and H7 in domestic poultry have either been stamped out or burnt out and disappeared. The current high-path H5N1 has been stamped out in Japan, South Korea, Thailand and many other countries in Asia, Africa and Europe—only to return again during the cooler months.

After the 1997 Hong Kong outbreak, lethal H5N1 viruses continued to spread south as far as Indonesia. To what extent that distribution reflected the (often illegal) movement of domestic birds rather than transmission by wildlife is not clear, though human transport was undoubtedly a factor. The fact that H5N1 has not jumped from Indonesia to north-western Australia can probably be attributed to the lack of any trade in poultry between those two regions, and to 'Wallace's line', which follows the deep-water channel along the interface between different tectonic plates. Running north to the Philippines, it effectively divides the fauna of some of the Indonesian islands and separates South-East Asia ecologically from Australia and New Guinea. This is undoubt-edly a key barrier that has helped to keep Australian wildlife and domestic species free of many exotic infections.

No such divide exists between Asia and Europe. It seems likely that the westward movement of high-path H5N1 viruses as far as Egypt, Scandinavia and the United Kingdom has largely been due to migrating waterfowl, though the distribution indicates that there is much we don't understand about those routes. Further research using radio tracking is currently under way. The first case of H5N1 viruses causing widespread fatal disease in wild birds occurred in 2002, when sparrows, pigeons and many aquatic species were found dead in Hong Kong's Penfold Park and Kowloon Park. Then, in 2005, a lethal outbreak killed more than 6000 migratory water-fowl at Qinghai Lake, a nature reserve that is a major breeding site in western China. This affected bar-headed geese, great black-headed gulls, ruddy shelducks and great cormorants. Other species that have since been found to be highly susceptible include whooper swans, black-necked cranes and pochards.

Long-distance westward spread of the Qinghai Lake strain was detected as early as 2005, the likely culprits being migrating ducks. Geese were exonerated because the infection was 100% lethal. Even in the mildly affected ducks, the virus changed from being a predominantly gastrointestinal-tract to a respiratory-tract infection. What had happened to make this high-path H5N1 virus so virulent for many species of wild birds and, incidentally, for mammals as represented by laboratory mice, ferrets, cats and us? The key modification was a single point mutation in one of the influenza polymerase (PB2) genes, again emphasising the fragility of the genetic relationship between these viruses and the bird species that maintain them in nature. If the infection had not been so much less severe in ducks, the disease would quickly have burnt out in wildlife and this virulent H5N1 virus would soon have disappeared.

Why this mutated pathogen emerged in wildlife living on Qinghai Lake, which is a protected, national reserve, was a little puzzling. In general, influenza A viruses maintain a reasonably amicable relationship with waterbirds living under natural conditions. A plausible theory is that the mutation occurred in a nearby intensive, domestic poultry operation and was then spread to the lake by bar-headed geese. Aiming to provide some variety in the diet of railroad workers in western China, the authorities initiated a commercial breeding program for bar-headed geese. Spread from chickens to geese could have been either direct or as a consequence of exposure to contaminated water sources. It only needed the escape of one recently infected goose, then a relatively short flight before symptoms set in, for the virus to reach Qinghai Lake.

The road to hell is paved with good intentions. As a consequence of initiatives by many agencies, including US and Australian aid organisations, the numbers of domestic poultry in Asia, particularly South-East Asia, have grown more than fiftyfold since the end of World War II. The aim has been to increase prosperity and to improve lives by developing local industries that provide affordable, high-quality protein to a rapidly expanding human population. The number of people on the planet has more than doubled since I graduated from the University of Queensland in 1962, with most of this growth being in poorer, developing countries. More mouths to feed also mean more pigs. House all these species at close quarters in a warm, wet environment and we have the ideal conditions for the emergence of a novel influenza A virus that will jump into humans.

One way to drive the emergence of mutant influenza viruses is to crowd large numbers of chickens together under conditions where there is rapid turnover. Even worse is the construction of 'high-rise' bird markets that bring many avian species together for differing times. Add the exposure of migrating birds to that equation and the situation becomes even more dangerous. That contact may not necessarily be direct. The waste from chicken houses is, for example, used to fertilise rice fields, a crop grown under the type of wet conditions that favour visits by waterfowl. The lesson is that if we are to change the basic ecology of natural systems, we must also be prepared to manage the consequences, informed by the underlying science. At least when it comes to birds in warmer, wetter climates, the influenza A viruses are an important factor to be considered. This isn't necessarily 'rocket science'. The solutions can lie with approaches as basic as proper waste management.

9

Bird flu guys

THE AVIAN INFLUENZA STORY didn't just happen, and it wasn't obvious. Common sense alone would never have led us to the conclusion that the influenza A viruses are maintained in nature by waterbirds. While the science of ornithology goes back several hundred years, the discovery of bird flu and all its ramifications belongs to the much younger discipline of virology. Ornithology is wonderful, not least because any committed person who is willing to commit time and pay the price of a good pair of binoculars, a field guide and a notebook can make a real contribution to studying bird numbers and migration patterns. That's pretty inexpensive, and the diligent watcher can function to great effect without having even the most basic science qualification. But so long as the influenza A viruses were living in reasonable harmony with their waterfowl maintaining hosts, observation alone could not have suggested their presence.

Working out an infectious disease requires years of in-depth training, costs a lot of money, demands specialised, high-tech equipment, and depends totally on sustaining major, high-quality research facilities like the Pasteur Institute, the Walter and Eliza Hall Institute, the Rockefeller Institute, CDC Atlanta, and the John Curtin School of Medical Research. Much of the necessary funding is inevitably supplied from the public purse, which is not surprising given that, apart from killing people, pandemics are both politically unpopular and economically disastrous. The global cost of the limited 2002 SARS outbreak, for instance, was in excess of US$50 billion.

Industry and some private charities, particularly the Rockefeller Foundation, the Wellcome Foundation in the United Kingdom and the Seattle-based Bill and Melinda Gates Foundation, also make major contributions to infectious disease research. The Rockefeller funded much of the early work on insect-borne viruses, and Wellcome does support some veterinary research, but it's not heavily into wildlife. Still, the question of where the influenza A viruses hang out in nature is so important for human medicine that there is no problem generating the necessary dollars via medical research budgets. The details of bird flu were established by scientists in academic environments—university medical schools and research institutes that are largely, though not exclusively, funded via the mechanism of competitive biomedical and (on a smaller scale) veterinary research grants.

Our understanding of avian influenza has evolved throughout my adult life. It's a fascinating story, and part of that narrative involves my two friends and colleagues Rob Webster and (the regrettably late) Graeme Laver, both of whom achieved the

accolade that has distinguished many top scientists from Isaac
Newton to Albert Einstein, the FRS—the Fellowship of the Royal
Society of London.

As an undergraduate, Graeme Laver was employed at the
Walter and Eliza Hall Institute as a humble laboratory technician while completing a part-time science degree at the University
of Melbourne across the street. He found himself in Alfred
Gottschalk's lab right when 'Uncle Alfie' was showing that the
Vibrio cholerae receptor destroying enzyme (RDE) functions by
splitting sialic (neuraminic) acid—an experience that was ultimately to define the major focus of Graeme's research career. In
1958, having completed his PhD in London, Laver was appointed
to the newly established Department of Microbiology at the John
Curtin School of Medical Research (JCSMR) in Canberra, headed
up by Frank Fenner, a trainee of FM Burnet.

Graeme Laver was a fearless, free and not always sympathetic
spirit. As Rob Webster says, 'Graeme was a man who would pour
oil on troubled waters, then set fire to it.' He lost a lot of friends over
the years, mainly because he couldn't tolerate any sort of authority
and was a major pain in the neck for many administrators. But he
also made the world a more lively and interesting place.

Rob Webster hails from the south island of New Zealand
and graduated in microbiology from the University of Otago,
Dunedin, which, then as now, had a very strong microbiology
program. He went on to work on influenza at St Jude Children's
Research Hospital (SJCRH), Memphis, where he was instru-
mental in recruiting me to work in the same area. My expertise is
in the host response, while he covers the virus aspect; it's been a
good synergy.

The Laver–Webster partnership began when they got together
in Canberra in a three-way interaction with Rob's PhD supervisor,
Stephen Fazekas. Basing his research career in Canberra, Graeme
generally stayed with the Websters when he visited Rob's SJCRH
laboratory. We often saw them together.

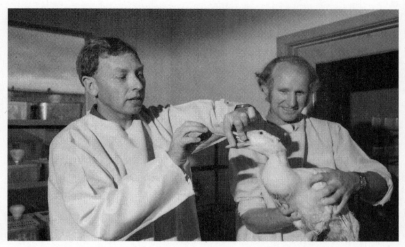

Graeme Laver (*left*) and Rob Webster. (Courtesy Multimedia JCSMR, ANU)

Through the years, Webster and Laver remained tightly focused on influenza and did a great deal together, including showing clearly that antibody drives antigenic drift, advancing our understanding of the genetics of virus reassortment, and developing the first 'split' influenza vaccine. And it was this quest to understand influenza in all its guises that eventually led them to look more specifically at bird flu.

Back in1962, an influenza A virus was found to be responsible for a major kill-off of South African terns, but there had been very little follow-up. Then Helio Pereira, the highly regarded Brazilian virologist who was, at that time, head of the WHO Influenza Center at Mill Hill, decided to take a closer at influenza in birds. He and Czech scientist Bela Tumova had analysed some 15 different bird-origin influenza viruses by the time they were joined by Rob Webster as a sabbatical visitor. The final paragraph of the little note the three of them co-authored for *Nature* reads: 'These findings pose a number of interesting questions regarding the possible relationships between human and animal influenza viruses. If the A/Turkey/Massachusetts strain had been isolated before

1957 it would have been tempting to suggest at that time that the human influenza A2 sub-type had been of avian origin.' The hunt was on. Could they find evidence that these viruses indeed cross from birds into humans?

Not long after, in the early 1960s, Graeme and Rob were walking together along one of the beaches near Canberra and were amazed by the large numbers of dead shearwaters (mutton-birds). Might such 'wrecks' be due to influenza? Could shearwaters be carrying influenza viruses during their long, annual migration between the Arctic and the coastline of Australia? Being scientists, there was no point in just speculating. They had to look.

Graeme asked his boss, Gordon Ada, to fund a field study from university resources. Predictably, Gordon told him to get lost. This was obviously another Laver stunt, and there was no way he'd allocate tax payers' dollars for some mad adventure that involved chasing after birds. Graeme then approached Martin Kaplan, who shared Pereira's interest in the possible significance of avian influenza and was at that time heading the WHO zoonoses program in Geneva (investigating infections that transmit from animal to man). Martin provided the massive sum of $500 for Graeme to develop a trapping and sampling project to survey the pattern of influenza A virus distribution in seabirds.

The short-tailed shearwater nests on the coastline and offshore islands of Bass Strait, the somewhat chilly stretch of rough water that pounds the beach near our house at Point Lonsdale near Melbourne. Doing fieldwork with this avian population is also made somewhat less attractive by the fact that they tend to share their burrows with the large, aggressive and very poisonous tiger snake. Fortunately for Graeme's plan, the wedge-tailed shearwaters (the short-tailed shearwaters' close relatives) don't all travel so far south. The focus of his seabird survey effort would thus be the wedge-tailed shearwaters, and other aquatic birds that frequent the much warmer and more attractive tropical islands of Australia's Great Barrier Reef. Graeme chose Tryon Island, about 50 miles off the Queensland coast, for what was to become an annual event that ranked very high on the list of 'Laver spectaculars'. Many in the flu community made the trip, which also involved a long, and sometimes very rough, boat ride.

Always thinking about the flu neuraminidase (N) that he'd first encountered as a junior technician with Gottschalk, Graeme used a neuraminidase inhibition assay to survey the mutton-birds for influenza virus–specific antibody. The scene for further expeditions was set in that first trip when he found that 18 of the 320 bird sera they'd collected were positive for antibodies to N2. In 1957, the H2N2 'Asian' flu virus had caused the second pandemic of the twentieth century, so it looked like they might just be onto something important. Over the years, they also isolated several influenza A viruses from different seabirds, including a wedge-tailed shearwater. It turned out, though, that the most important was an H9N9 virus recovered from a black noddy tern.

That H9N9 bird virus was to provide the basis for what was, at least in practical terms, Graeme Laver's greatest scientific achievement. Early in his career, Graeme had learnt to grow crystals. At the interface of biochemistry and physics is the science we now call structural biology. Crystals of purified molecules are bombarded with intense radiation—X-rays in the early days but now the beamlines from a cyclic particle accelerator (a synchrotron). Recording the displacement of the X-ray or electron beams produces diffraction patterns that are then 'solved' by very smart, highly trained people to give the three-dimensional structure of, say, DNA or protein. Now computers are a big help, but early on, a lot of the 'visualisation' that led to the final construct was done in people's heads.

Back in the mid-1970s, Graeme managed to produce some rather suboptimal, flat crystals of N2 that were, nonetheless, good enough to allow the X-ray crystallographer Peter Colman and his Melbourne CSIRO Division of Protein Chemistry team to determine the basic structure of the flu neuraminidase. But it was the N9 from their Great Barrier Reef noddy tern virus that yielded the most spectacular crystals.

Working out what the influenza neuraminidase actually looks like was fascinating in the scientific sense, but it also had practical value. Knowing the structure allowed them to identify the molecular region that cleaves the sialic acid bond and allows newly made virus to be released from the dying cell. Colman got together with a young carbohydrate chemist, Mark von Itzstein, who was just down the street at the Victorian College of Pharmacy. Mark used computer-modelling approaches to design a small molecule mimic, a drug that blocks influenza-virus neuraminidase function by binding to the active site. The result was the first specific anti-influenza treatment, zanamivir, later marketed as Relenza. Relenza was also one of the very first therapeutics produced by combining structural biology analysis and computer simulation, a process that is basic to rational drug design, the Holy Grail for the pharmaceutical industry.

Meanwhile, as Laver continued to focus much of his scientific effort on neuraminidase biochemistry, Webster progressively built a global avian influenza A virus-monitoring network. Influenza virus surveillance involves bleeding infected subjects to provide serum for antibody analysis and, more importantly, sampling mucosal sites to look for infectious virus or its footprints. That latter process has become much simpler with contemporary PCR techniques. With humans, the standard approach is to swab the nose. But influenza is normally a gastrointestinal tract infection in birds, so that involves taking a cloacal swab. Basically, Webster and his colleagues have spent a good part of their lives sticking probes up the back ends of birds across the planet, an achievement that has, at one time or another, been the cause of some low hilarity.

From the Russian tundra to Antarctica, Rob has roamed the globe chasing avian influenza viruses. Graeme would sometimes tag along, especially if the expedition was to some outlandish place. Even the rather discriminating and theatrical Allan Granoff,

Rob's close Memphis colleague, was persuaded to accompany them to the Guano Islands, off Peru. Guano is comprised of bird droppings that have accumulated over the millennia. Thoroughly permeated with Guano dust, Alan proposed for a time that he might as well change his name to 'Guanoff'.

Now in his late seventies, Rob Webster recently retired, or so he claims. So far as I can see, he's still in great shape and flies regularly around the world, consulting with those working on avian influenza and advising governments and regulatory authorities on what they should do to minimise the impact of bird flu.

Not realising, or perhaps not acknowledging, that he had fatal cancer, Graeme Laver collapsed on a flight over Moscow en route to an influenza meeting in Portugal. The traffic control authorities cleared all airspace to London so that he could receive medical attention as soon as possible. That was Graeme Laver: even his exit had to be spectacular and cause maximum disruption! We all miss him.

10

Bug detectives

SO FAR, WE'VE TALKED a lot about viruses, but we've largely ignored that microscopic universe of other, more complex infectious organisms that cause severe disease problems in birds, and in mammals like us. Though viruses undoubtedly belong in 'bug world', virologists like me tend to use the word 'bug' to describe the plethora of bacteria, fungi and parasites that (unlike viruses) do not necessarily depend on using the internal machinery of some host cell to survive in nature. While influenza has only eight genes, the biggest viruses (poxviruses) have around 250 and are almost as complex as the smallest bacteria (mycoplasma, with more than 450). The chlamydia that cause trachoma and psittacosis have more than 900 genes, compared with 2000 plus for *Neisseria meningitidis* (meningitis in kids), 4000 or so for the various mycobacteria that cause tuberculosis in a range of species, and more than 5000 for the different malaria protozoa.

Smeared onto glass slides and stained with some appropriate chemical dye for light microscopic analysis, these larger parasites are readily seen by an experienced observer. This made them much easier to study early on than the viruses, which can only be visualised with the much greater power of electron microscopy. Focusing on these bigger bugs thus allows us to linger a little on the relatively short history of infectious disease research and to tell a couple of stories of how studies in birds contributed mightily to the revolution in scientific understanding that occurred during the last half of the nineteenth century.

* * *

It's sometimes hard for us to relate to the way people thought in earlier times, but it's a simple fact that many who contributed so much to the ways that we still experience the world—from Homer, Saint Paul and Will Shakespeare to Mozart and Jane Austen—lived in complete ignorance of the myriad microorganisms that live in or on every vertebrate species. The body weight of an adult human is, for example, augmented by at least 1–2 kilograms of normally harmless gut bacteria, which both provide some essential nutrients for us and take up 'space' in ways that inhibit the growth of potentially dangerous bugs in the crypts and crevices of the gastrointestinal tract.

But it's only those parasites and other contaminating organisms that can't be seen with the naked eye that escaped the attention our ancestors. As voyages lengthened from weeks to months, every eighteenth-century sailor in Britain's Royal Navy became all too familiar with the squirming weevils in their 'hard tack' biscuits and salt pork. The ancient Egyptians described the round worms and tapeworms that grow in our gastrointestinal tracts (and in those of our avian cousins). Not surprisingly, they were also aware of the female guinea worms that can horrifyingly emerge from a human skin ulcer. Then there are the various insects that, like worms, infest rather than infect us. It's hard to miss the fact when, after walking in a grassy and/or wet landscape, we find ourselves the unwitting host of a tick or a leech.

Some parasitic insects even attracted the attention of the great poets. Robert Burns' 'To a Louse' records an encounter that was probably more frequent in the eighteenth century than it is today. And the common house fly was memorialised by the religious dissenter William Blake (1757–1827) in his pre-dieldrin insect poem 'The Fly':

Then am I
A happy fly,
If I live,
Or if I die.

Blake did not realise, of course, that his 'happy fly' carried bacteria that contaminate exposed food.

His younger contemporaries, the romantic poets John Keats (1795–1821) and Percy Bysshe Shelley (1792–1822) also knew nothing about infectious microorganisms. Both wrote fondly about insects, but to celebrate burgeoning life in the spring rather than from the infestation point of view. Even so, it's been said that they, and their friend Coleridge, were the last poets who knew anything much of science. Nevertheless, when Shelley wrote, 'I weep for Adonais—he is dead', beginning his long poem mourning the death of the 25-year-old John Keats, he had no understanding of what had caused the disease that killed his friend. At age 15, Keats lost his mother from consumption, the chronic, debilitating lung disease we now call tuberculosis (TB). Then he would have been even more heavily exposed to the tubercle bacillus when he nursed his brother Tom. Tom Keats died in 1818, and in 1821,

John Keats succumbed to the same infection. The following year, Shelley drowned in a boating accident in Italy. Though the causes were different, these two romantics both departed because their lungs had filled with fluid, they could no longer breathe and there was no oxygen for their red blood cells to carry to what were undoubtedly superior brains.

The world then waited for more than six decades before, in 1882, the great German microbiologist Robert Koch reported his triumphant discovery of *Mycobacterium tuberculosis*, the cause of human TB. Human beings have been around for somewhere between 120 000 and 200 000 years, and we're going back less than two centuries when we think of Keats and Shelley. Masters of the English language, they would have been completely unfamiliar with the terms 'microbiologist', 'bacteria', 'genetics', 'tubercle' and 'DNA'. Shelley could have had no idea that the skylark immortalised in his famous poem might die in much the same way as his friend Keats, as a consequence of exposure to the closely related *Mycobacterium avium*. This organism has been shown to cause TB-like pathology in many species of birds (including Australian cassowaries) and can be passed on to us if we are immunosuppressed—by AIDS, age, or treatment with cytotoxic drugs (for cancer). The psittacines are particularly at risk, and of course, some of us live very close to parrots.

People had some idea of contagion before the middle of the nineteenth century, but they made no intellectual connection between infestation with visible parasites like worms and ticks and the unseen world of infection with microbes. Rather, their perceptions were shaped by the idea of 'miasmas', dangerous fogs that arose in wet marshlands, for example, and carried the risk of fever and death. It's easy to see the origins of that line of thought in waterborne bacterial diseases such as cholera and typhoid, or in mosquito-transmitted infections like malaria and flavivirus encephalitis. Though (as with anthropogenic climate change) it took decades to convince those who would not examine the evidence and who adhered rigidly to other explanations, miasmas were finally eliminated as a cause of disease by Louis Pasteur's experiments in the early 1860s, which disproved the idea of spontaneous generation and established that infectious microorganisms are independent, transmissible entities.

Building on Pasteur's discovery, Robert Koch then laid out what have come to be known as Koch's postulates:

- The microorganism must be found in abundance in all organisms suffering from the disease, but should not be found in healthy animals.
- The microorganism must be isolated from a diseased organism and grown in pure culture.
- The cultured microorganism should cause disease when introduced into a healthy organism.
- The microorganism must be re-isolated from the inoculated, diseased experimental host and identified as being identical to the original specific causative agent.

Providing we set aside situations like HIV/AIDS—where the destruction of the human immune system leads to the invasion of bugs (like *cryptosporidium* and *pneumocystis*) that normally live benignly on the mucosal surfaces of the gut or respiratory tract or, like *M. avium*, are readily controlled if we encounter them in nature—Koch's postulates are still true, and will have been satisfied for all of the infections of birds discussed in this book.

Apart from his massive contribution to the understanding of human disease, Pasteur also has the distinction of being, perhaps, the very first bird-bug detective. He studied avian cholera, a highly lethal bacterial infection of both domestic chickens and wild birds that can kill as quickly as high-path avian influenza (Chapter 8). After many years of failure, Pasteur and his team worked out how to grow the fowl cholera agent, now called *Pasteurella multocida*, in broth cultures. Then some of the cultures just sat around for a month when everyone in the lab went off on their summer holidays. On injection, these 'aged' cultures failed to fulfil Koch's postulates and no longer caused severe disease. The big discovery came when the surviving chickens were found to resist a second challenge with virulent *P. multocida*. Pasteur and his young colleague Charles Chamberland had discovered by chance that 'weakening', or attenuating, an infectious agent can lead to the development of a protective vaccine. He later applied that approach to anthrax, and various types of attenuation (often serial passage through cell cultures, chick embryos or laboratory mice) have since become basic to the development of most current live vaccines, including those that protect us against poliomyelitis, measles and yellow fever.

After Edward Jenner's 1796 discovery that inoculation with cowpox protects against subsequent challenge with smallpox virus, Pasteur and Chamberland's work with chicken cholera in 1880 was the first real study of vaccination. It's extraordinary that Jenner—following the example of those in the Islamic world, and in China before that, who practised 'variolation' by scratching fully virulent smallpox virus into the arms of children as a protective measure—operated in complete ignorance of how infection works. Pasteur, on the other hand, knew what he was doing and why, as have all those who've followed.

Pasteur's influence was enormous. By 1886, his student, the veterinarian Edmond Nocard, had cultured *M. avium* from birds. In 1888, Nocard also discovered *Nocardia asteroides*, an organism that causes a TB-like granulomatous condition. Though it's a relatively uncommon cause of avian disease, it has been recovered from sick parrots, crakes and some other species. Then, in the following decade, he and another Pasteur Institute scientist, Emile Roux, were the first to identify the *Mycoplasma* that can induce respiratory symptoms in most terrestrial vertebrates, sometimes called 'walking disease' in humans as we generally shrug off what we regard as a mild cold and do not take to our beds or the recliner in front of the TV. The avian pathogen *Mycoplasma gallisepticum* causes severe sinusitis in turkeys and slow-onset, chronic respiratory disease in chickens, pigeons and other wild birds. Such infections may, for instance, be particularly problematic for migratory species, since the margin between survival and death can be very slim for these long-haul travellers.

In the latter half of the nineteenth century, once Pasteur and Koch had established the basic principle of infection, the new technologies that were soon developed enabled a process of rapid discovery. The capacity of Pasteur's nutrient broths (or soups) to support the isolation of microorganisms from diseased organs was improved by adding various 'growth factors' and by modifying the culture conditions. They discovered that some microorganisms require the presence of O_2 (aerobes) while others (anaerobes) do not. Unsurprisingly, as they grow in the lung, *M. tuberculosis*, *M. avium*, and *N. asteroides* are obligate (absolute) aerobes, while many of the bacteria that infect the vertebrate gastrointestinal tract are anaerobes. Microbes and, indeed, vertebrate cells from chick embryos and other sources might be cultured in Roux flasks, named after another eminent 'Pasteurian'. Those who joined the game early on enjoyed extensive naming rights. Pasteurisation is another example.

In case you've got the impression that all the early work on important bacteria was done in Paris or Berlin, it's worth pointing out that an American, Theobald Smith, discovered one of the major causes of enteric infections. Working at the newly formed Bureau of Animal Industry (BAI) in the US Department of Agriculture, Smith and his boss, Daniel E Salmon, authored an 1886 report that described the isolation of the organism we now call *Salmonella enterica*. Though the bug is named after Salmon, the credit for the discovery goes to Smith.

Theobald Smith's parents were German, and he was fluent in the language. Though most major research papers are now published in English, that wasn't true before World War II, when there was still a rich scientific literature in French and German. Being able to read German allowed Smith to access the work of Koch and his colleagues, and he soon applied the techniques they described to his work at the BAI. Since then, microbiologists have identified thousands of *S. enterica* variants, including *S. enterica typhi*, the cause of human typhoid.

Salmonella species are also a problem for avian species, with bird feeders being a prominent cause of spread to wildlife. The disease in poultry is often chronic, with the result that eggshells are commonly contaminated. Then, if there is even a slight crack, the bacterium can gain access to the rich protein environment that (in a fertilised egg) would nourish the developing chick embryo. The result is that salmonellosis is a very common cause of human food poisoning, particularly in situations where the eggs have not been properly washed, come from contaminated premises (as occurred in 2010 in Iowa) or are shelved under warm conditions that allow rapid bacterial growth.

Theobald Smith also speculated about the possible role of mosquitoes in transmitting malaria, the parasite that had been discovered in 1880 by the French scientist Alphonse Laveran. Laveran and the prominent English researcher Patrick Manson, who was the first to confirm that the cause of malaria had indeed been found, reached much the same conclusion, though, like Smith, they had no understanding of the malaria–mosquito interface. That was to be worked out a bit later by Ronald Ross.

Laveran was awarded the 1907 Nobel Prize, and Manson enjoys perpetual fame as a consequence of having an invasive worm, *Schistosoma mansoni*, named after him. This little horror lays its eggs in the liver, spleen and intestines. Masses of inflammatory white blood cells then surround the egg, leading to space-occupying granulomas (hard lumps), which cause severe clinical debility and even death. Water birds are also infected with different schistosomes, which spend part of their life cycle in freshwater snails and cause a much less severe human problem, swimmer's itch. This results from the invasion of a larval stage, the free-swimming cercariae, into human skin. Immersion in rice fields or muddy ponds can lead to a week or two of discomfort caused by the reaction that gets rid of these 'foreign bodies'. This avian schistosome itch led to a ban on pond-dunking during 'fresher' (first-year) initiation ceremonies at the subtropical University of Queensland, which I attended as an undergraduate. Fortunately, the avian schistosomes don't mature in us. The cercariae are soon eliminated by the inflammatory response, and the urge to scratch goes away. The failure to get rid of the much tougher *S. mansoni* eggs, though, is what causes the symptoms of intestinal schistosomiasis, which affects, with varying severity, some 200 million human beings in tropical countries.

Ronald Ross was born into an expatriate British medical family in India, qualified in London and worked for a time with Manson. Appointed to Secunderabad as a young army medical officer in 1895, Ross bred mosquitoes from larvae, allowed them to feed on a malaria patient and then trapped the blood-engorged insects for later dissection. Once you know what to look for, the various life forms of malaria can readily be seen down the microscope. Following on the earlier studies of Laveran and Manson, Ross was able to trace the escape of these protozoa from human blood into the mosquito. Continuing his experiments in Bangalore, he also looked at the idea (suggested by Manson) that people contract malaria when they drink water contaminated by dead, infected mosquitoes. After considerable effort, Ross could generate no support for the waterborne hypothesis, which is, of course, right for bird schistosomes and wrong for all forms of malaria.

Back in Secunderabad in 1897, Ross continued his experimental studies with mosquitoes fed on infected patients and made his first major discovery. After many careful and exhaustive dissections, he found malaria parasites multiplying in the stomach wall of the mosquito. Ross was delighted. The part played by the mosquito in malaria transmission was staring him in the face. It must transmit when the mosquito bites and takes a blood meal. All that he needed now was to do the necessary experiments.

Then disaster struck, or so he first thought. Ross was transferred to Calcutta, where the incidence of malaria was low. He was a serving army officer, and in any case, there were, at that time, very few jobs where it was possible to pursue any form of scientific research. Ross couldn't just throw a 'hissy fit', resign and head off to UCLA or Harvard. Then he remembered that birds also get malaria. As he recalled in his 1902 Nobel Prize lecture, 'a number of crows, pigeons, weaver birds, sparrows and larks were then immediately procured, and experiments commenced on them without delay'.

Ross was working with two different species of malaria-like parasites: one he called *Halteridium* (now classified as *Haemoproteus*) and the other *Proteosoma* (now *Plasmodium*). Looking down the microscope, he could tell the difference between them, and he soon showed that the grey mosquitoes (*Anopheles*) became infected when exposed to sparrows that were circulating *Proteosoma* (but not *Halteridium*) and would later transmit that same parasite to clean sparrows. Using the avian malaria model, Ross established that, though their life cycles were ultimately found to be identical, the malaria parasites of humans and birds are distinct. Furthermore, he established that these two strains of protozoa are carried by different species of mosquitoes, and proved beyond any doubt that malaria transmits when an infected mosquito takes a vertebrate blood meal.

Ronald Ross is an interesting character. He also painted, and he wrote and published plays, novels and poetry. In 1897, he penned the following to celebrate his first big discovery:

> *This day designing God*
> *Hath put into my hand*
> *A wondrous thing. An God*
> *Be praised. At his command*
> *I have found thy secret deed.*
> *Oh million mudering Death, I know that this little thing*
> *A million men will save*
> *Oh death where is thy sting? Thy victory oh grave?*

Well, he's no Keats or Shelley, but he was a great scientist (and a Scot, which explains the 'mudering' rather than 'murdering'). Ross was awarded the second Nobel Prize ever for medicine—though clearly not for literature. The first medical Nobel went (in 1901) to Emil von Behring, who had worked early on with Robert Koch. Koch was recognised in 1905 for discovering the tubercle bacillus, and Pasteur could have won several Nobel Prizes, but he died too soon, in 1895. Much of his best work, including the *P. multocida* vaccine discovery, was done after he turned 45 and suffered a severe stroke, which left him paralysed on the left side for a time and with a permanent limp. These guys were tough birds, and birds are part of their story.

11

Hawaiian wipeout

HAWAII IS RIGHTLY REGARDED as the birthplace of surfing, and every surfer is all too familiar with the reality of wiping out. The wipeout we're talking about here, though, isn't to do with tube waves, beach breaks and being dumped head first into the sand at Lanikai or Hanalei Bay, but with the disastrous effect of malaria on Hawaiian bird species.

The immediate reaction of a medical infectious disease expert to this might be to declare that there's no malaria problem in Hawaii. And that's true if we're talking only about the human disease caused by parasites like *Plasmodium falciparum* or *Plasmodium vivax*, but it's certainly not the case when we extend our concern to the birds of the Hawaiian Islands. Pathogens like

West Nile virus are potentially lethal for corvids, raptors and us (as we saw in Chapter 5), and broader community awareness of such infections is greatly heightened by the fear that accompanies our potential vulnerability. But birds also have their own parasites that may cause disastrous loss of life in particular avian species without putting humans at direct risk.

Any discussion of malaria in Hawaii, then, focuses on the catastrophic wipeout of wild birds caused by the protozoan parasite *Plasmodium relictum*. Carter Atkinson and his colleagues described what happens when this parasite infects Amakihi, one of the many species of honeycreeper native to the Hawaiian Islands. These particular birds were trapped at cool, high elevations, where there are no mosquitoes, then brought down to sea level:

> Mortality among Amakihi exposed to a single infective mosquito bite was 65%. All infected birds had significant declines in food consumption and a corresponding loss in body weight over 60 days. Gross and microscopic lesions in birds that succumbed included enlargement and discoloration of the spleen and liver and parasitemias (infected cells in blood) as high as 50% of circulating erythrocytes (red blood cells).

This account of avian pathology also fits well for the commoner form of human malaria.

In addition, symptoms of fever, suppressed appetite and weight loss are characteristic for humans and corvids infected with West Nile virus. Like malaria, yellow fever is associated with severe liver pathology, but birds aren't a natural host for yellow fever virus, and the damage caused by malaria reflects clogging of the microcirculation with parasitised erythrocytes rather than the direct infection of hepatocytes (liver cells). That 'mechanical blockage' effect can also be associated with fatal neurological symptoms in all the vertebrate species that have their own malaria parasites.

When it comes to disease processes, the furry and feathered members of our planetary biota are affected and respond in very similar ways. Once an incision has been made, the organs are exposed to view and the tissues are subject to microscopic analysis; the organs of dead people, birds and laboratory mice are, though obviously varying in size, not so different to the trained eye of the pathologist. Depending on your point of view, this similarity might reflect a shared evolutionary history, the fact that we're all God's creatures or both. Where terrible pathogens like *Plasmodium*, yellow fever virus and West Nile virus fit into any non-Darwinian view of the world is, however, a bit harder to fathom, at least in the context of an amiable and caring deity.

I won't go into the details of the complex malaria life cycle, but what happens first from the point of view of the vertebrate host (bird or human) is that a form of the malaria bug gets into the blood stream when injected in the saliva of a biting mosquito. This is followed by a multiplication stage in the liver and the release of more parasites into the blood. These invade the circulating red blood cells (erythrocytes), causing the parasitemia that in turn provides the source of infection for any virgin mosquito taking a blood meal. Diagnosing malaria is easy. Blood smears are stained with simple chemical dyes like Giemsa, allowing the intraerythrocytic black dots of the parasite to be readily seen using the type of microscope found in any undergraduate biology laboratory. The progress and severity of the infection can be measured by simply counting the number of red blood cells that do, or do not, contain black dots.

It's possible to recall an Elysian era when avian malaria was not an issue in Hawaii, but we would need to go back almost 200 years, to a time before Louis Pasteur (in Paris) and Robert Koch (in Berlin) had defined the germ theory of infectious disease. While the 1999 West Nile virus epidemic was illuminated from the outset by the enormous power of contemporary molecular science, an understanding of what happened early on in Hawaii depends essentially on historical reconstruction from the careful observations made by early naturalists and ornithologists.

Organised science has been around for longer than most of us realise, and European academies such as the Royal Society of London were already sending researchers out to Hawaii in the latter part of the nineteenth century to measure and record what was happening with this unique island biota. They played by the same basic rules of science that we follow now, though the technologies have dramatically changed. While there were many fewer professional scientists back then, their lives were as much governed by hypothesis, experiment, observation and careful measurement, then peer-reviewed publication in well-regarded journals, as is any academic ornithologist (or climate scientist) today. The *Philosophical Transactions of the Royal Society of London* began publication in 1665 and is still in print, after spawning other journals including the *Proceedings of the Royal Society* (from 1854), which publishes a lot of wildlife research in its 'B-for-biology' series. The top British science journal *Nature* first appeared in 1869.

While many have worked on the problem more recently, the seminal account of the Hawaiian avian malaria problem is a 1968 review in *The Condor* by Richard Warner of the Museum of Vertebrate Biology at the University of California, Berkeley. Warner's intellectual synthesis was that, though migrating birds carrying *P. relictum* may well have landed in Hawaii over the millennia, the disease did not spread because of the absence of a suitable mosquito vector.

That situation is thought to have changed in 1826, when, according to the Reverend William Richards, a watering party from the ship *Wellington* discarded 'dregs alive with wrigglers into a pure stream, and thereby to blot one more blessing to the Hawaii that had been Eden'. In fact, the loss of paradise had begun much earlier, with James Cook's third voyage of exploration, which ended tragically in the death of the 'great navigator' at Kealakekua Bay. The 1778 arrival of HMS *Resolution* and HMS

Discovery marks the beginning of Hawaiian colonisation by foreign biota, including European humans. The 'wrigglers' of William Richard's account almost 50 years later were likely the larval form of the malaria vector *Culex quinquefasciatus*. Writing long before the time of Ronald Ross, he was, of course, commenting only on the discomforting whine and associated mosquito bites, not avian malaria. That connection was yet to be made.

Richards himself was a missionary bringing, of course, another type of transforming invasion. And malaria wasn't the only avian disease problem carried to Hawaii from other shores. Writing in 1903, the naturalist Robert Perkins describes 'bumble foot', or bird pox (avipox), in lowland populations. Again, like the malaria parasites, the poxviruses tend to be very specific to a particular vertebrate host. However, avipox differs from *Plasmodium* in that this big virus (around 200 genes) is very resistant to environmental degradation, and while it can be transmitted mechanically by biting flies and mosquitoes, it doesn't replicate within the insects and is also spread by bird-to-bird contact. Thus, while the distribution of *P. relictum* malaria is determined by the range of *C. quinquefasciatus*, avipox can potentially spread whenever infected and normal birds are brought together.

The fact that Hawaii is comprised of eight large and ten smaller islands meant that not all areas were affected simultaneously by either avipox or malaria. This provided those early scientists with an extraordinary opportunity to observe the spread and assess the effect of these two diseases on native birds. Warner relates that, while bird populations on some other islands remained relatively intact, six of the 11 passerine species on Oahu were extinct by 1900, even though they were not hunted for their feathers, and their habitat remained undisturbed. He quotes WA Bryan, who in 1915 sadly observed that 'Oahu can make the melancholy boast that it has a greater list of extinct birds, in proportion to the total number of species from the island, than any other like area in the world.'

Warner also describes his own experiments, conducted in 1958. Aware that neither bird pox nor *C. quinquefasciatus* had reached the remote, leeward Hawaiian island of Laysan, he brought 24 Laysan finches to Honolulu, transporting them securely aboard a US Coast Guard vessel in cages wrapped in several layers of fine-mesh but porous cheesecloth. The birds lived happily and ate well for two months while enclosed in a screened room and caged in their protective cheesecloth. Then the covers were removed and the windows opened, and within two weeks, they saw the first characteristic swellings of bird pox, followed by the inexorable spread of the disease through the experimental group.

The fact that avian malaria did not emerge in this first experiment was thought to reflect the fact that *Culex* numbers in Honolulu were extremely low at that time. Warner thus repeated the study in the following year. This time the birds were brought to Lihue on Kauai. The protected controls remained in good health, while those that were placed outside in unscreened cages all developed evidence of massive *Plasmodium* parasitemia and were dead within 16 days.

When I summarised this avian malaria story to a friend over coffee, the immediate response was, 'But birds are covered in feathers. How do mosquitoes get to bite them?' Well, Warner covers that too. A feature of *C. quinquefasciatus* that I've neglected to mention is that it's a night flying mosquito. When sleeping, the relatively resistant Japanese white-eye (a non-native Hawaiian) tucks its bill and face back into fluffed back feathers, fluffs its breast feathers and crouches down so that its belly touches the perch. This minimises access to the corner of the bill, the forehead and the legs and feet, all regions that remain exposed as possible mosquito feeding sites on sleeping, susceptible birds. As with many infectious diseases, behaviour may not be everything when it comes to vulnerability, but it is important.

Estimates suggest that approximately half of the 140 native Hawaiian bird species are now extinct, with another 30 classified as endangered and 12 of those at such low numbers that they are on the brink of disappearing. While malaria and avipox are very likely to have been significant contributors to this massive loss of diversity, other factors will also have been at work. Hunting, degradation of habitat, attack by cats, dogs and rodents, and competition with introduced birds are all obvious risk factors. In addition, these different influences might be expected to work together. The reduced mobility associated with 'bumble foot', or the lethargy that results from fever and not eating as a consequence of malaria, would likely make the infected bird much more vulnerable to predators.

We get an idea of the extent of the damage caused by mosquito-born diseases by looking at the characteristics of honeycreeper species that live at different altitudes throughout the Hawaiian Islands. Some species will move down the mountains as the flowers that provide nectar bloom first at lower levels, then back up again with the continued progress of warming from spring to summer. Groups of other honeycreepers, like the Amakihi, tend to be less mobile and prefer to stay within a relatively small altitude range. This protects the Amakihi living above the *C. quinquefasciatus* line at about 1400 metres, though birds at the base of the mountain should be fully susceptible. One encouraging sign is that the numbers of Amakihi living satisfactorily at low altitudes seem to be increasing. Many have low levels of malaria parasitemia, a characteristic of other more resistant bird species that generally survive this infection.

What is even more exciting in the scientific sense is that geneticists are finding substantial molecular differences in Amakihi populations sampled from low, intermediate and high altitudes. This fits with the classical Darwinian idea that contracting these mosquito-born infections has, over time, led to the selection of a more 'fit' species, which can now 'tolerate' or control these disease processes. It's possible, then, that natural selection—operating over the century or more since avian malaria and poxviruses first invaded Hawaii—has now led to the emergence of resistant sub-strains, offering hope for the future.

A very real concern is that the high-altitude, mosquito-free zones in Hawaii will essentially disappear with the 2°C or more increase in ambient temperature that climate scientists are convinced will inevitably occur before the end of the twenty-first century. Clearly that development will be less catastrophic for those species that have been able to evolve in response to the infectious disease challenge. Such processes must have happened over and over through geological time, though this time the greenhouse effect, driven by our fossil fuel addiction, is pushing the rate of environmental change to fast-forward. As any palaeoecologist will tell you, most of the species that have ever existed are no longer around. We are currently living through the sixth great extinction in the history of this planet. Perhaps we might think about the possibility of selectively breeding some species of wildlife for resistance to insect-borne pathogens, the diseases that are likely to become more problematic with anthropogenic climate change.

12

The great parrot panic of 1929–30

IT SAYS SOMETHING PROFOUND about the human condition
that history takes much greater account of wars and military
mayhem than of the enormous losses caused by communicable
disease. Infection accounted for more than 60% of the deaths in
the American Civil War of 1861–65, while many more died from
influenza in 1918–19 than were sacrificed to explosive shells, bombs
and bullets in that idiot tragedy we call the 'Great' War of 1914–
18. The literature on World War I is massive, but I know of only
one major work from the inter-war (1919–39) period, the novella
Pale Horse, Pale Rider (1939) by Katherine Anne Porter, that is at
all focused on the 1918–19 influenza catastrophe. Until the bird
flu scare of the 1990s, the single non-fiction account for a popular

audience was Alfred W Cosby's *America's Forgotten Pandemic: The Influenza of 1918*, which is in many ways still the best book on the subject. Throughout human history, and particularly since we started to gather ourselves and our domesticated animals and birds into agricultural communities, infection has caused many more deaths than conflict. But it seems that we much prefer to deal with realities where there are conniving, identifiable human bad guys than to contemplate a universe where we are at the mercy of mindless microbes and nature. Both are brought together in Dennis Lehane's recent novel *The Given Day*, which interweaves the 1918 influenza epidemic, baseball, Babe Ruth, racism, politics and the confrontation between good and evil during the Boston police strike of that time.

Similarly, when we hear someone refer to the panic of 1929–30, our thoughts go immediately to the stock market crash that caused a great deal of human suffering and some suicides. But these years are also marked by other deaths associated with a quite different kind of panic. The story was summarised in the *New Yorker* on 1 June 2009. Jill Lepore's *It's Spreading: Outbreaks, Media Scares and the Parrot Panic of 1930* tells a story that is in many ways reminiscent of what happened during the 2002 SARS outbreak, though the later event was much more serious.

The 'parrot panic' story was new to me, though it was evidently written up in Paul de Kruif's *Men against Death* (1932), the sequel to his more famous *Microbe Hunters*. By the late 1920s, popular writers such as de Kruif and Sinclair Lewis had contributed significantly to increased public consciousness of the dangers of infectious disease. People worried about 'germs'—an obsession that may have sent Howard Hughes completely mad—and contamination became a productive focus for advertising agencies and manufactures of products like *Lysol* (carbolic acid).

What Lepore describes in her *New Yorker* article is how three cases of severe psittacosis in a prominent Baltimore family, together with the death of their recently purchased pet parrot, escalated into a media-fed panic. I won't attempt to reprise her elegant and informative analysis, but she identifies articles in the *Washington Post*, the Chicago *Daily Tribune*, the *New York Times* and the *San Francisco Chronicle* as fuelling the sense of public concern. Other contemporary accounts in *The New Yorker* and in *Life* magazine were evidently more sceptical. When it was all over, there had been less than 140 human clinical cases and many fewer deaths, though enormous numbers of psittacines of one sort or another paid the ultimate price.

The job of sorting out the 1929–30 psittacosis outbreak was given to medical doctor Charles Armstrong and his technical assistant Henry (Shorty) Anderson of the US Public Health Service's Hygienic Laboratory in Washington, DC. Armstrong caught the infection himself and recovered, but Shorty Anderson, Dr Daniel S Hatfield (the head of the Baltimore City Health Department) and his colleague Dr William Stokes all died. Clearly, these men weren't just office bureaucrats but 'hands-on' public health officials. Like SARS, the disease was much more dangerous in older people, and all those who died were more than 30 years of age—a somewhat sexist name for psittacosis was 'old maid's pneumonia'. As Armstrong relates, the incidence of human infection was higher in women, who spent more time at home and also cared for pet birds in typical 1930s American households.

Armstrong published his own short account of what happened in the *American Journal of Nursing* in 1933. He points to the first description of psittacosis, including the link to infection by birds, by a Dr Ritter in Switzerland. The disease was named psittacosis in a description of an 1892 Paris outbreak, and in 1896, Edmond Nocard (whom we met in Chapter 10) thought he had isolated the causative bacterium, but he got it wrong. The problem was the inadequacy of the then available technology. Much the same mistake was made during the course of the influenza pandemic 20 years later.

Psittacosis is caused by *Chlamydia psittaci*, an organism that, though a bacterium, shares with viruses the characteristic of being an obligate intracellular parasite. Small enough to pass through larger filters that exclude bacteria capable of growing in cell-free culture media, it would be held back by, for instance, pore sizes that allow poliovirus, influenza virus or yellow fever virus through. Again like the viruses, much initial research on the *Chlamydia* depended on the fact that they grow in the membranes of embryonated hens' eggs.

Armstrong and Anderson showed that psittacosis is generally a mild, clinically inapparent infection of parrots, which can be long-term, healthy carriers. Unlike the situation with SARS, they did not see a single case of the disease spreading between humans. Nowadays, psittacosis is readily treated with tetracycline and doxycycline, so it's important that anyone who develops a high fever should alert their doctor if they've been spending a lot of time with a parrot.

Though at least 33 Americans died from psittacosis in 1929–30, respiratory infections (including influenza) were thought to account for more than 10% of all US deaths in that year. 'Parrot disease' was thus a fairly small player in the overall spectrum of infections. Indeed, the acerbic commentator EB White described 'parrot fever' as 'the latest and most amusing example of the national hypochondria'. Still, as the eminent Swiss-born, California-based microbiologist Karl Meyer observed several years later, the much-publicised events of 1929–30 served to improve the monitoring and general wellbeing of captive psittacines, and to alert public health authorities throughout the world.

Beyond that, though, the 1929–30 'parrot panic' had one very significant impact. On 26 May 1930, the US Congress expanded Washington's Hygienic Laboratory and renamed it the US National Institutes of Health—the NIH that now funds more public-sector biomedical science globally than any other organisation. Though small when compared with current levels of military spending, the 2009 NIH budget was in excess of US$29 billion. Operations like, for example, the influenza virus-watch program at the University of Hong Kong (discussed in Chapter 8) have received substantial funding from NIH sources. So when it comes to the health of humanity, those 'sentinel psittacines' that gave their lives in 1929–30 ultimately contributed to a movement towards scientific medicine that has been of immense human benefit and that has also helped in the control of avian disease.

13

Catching cancer

VIRUSES, CHICKENS AND CANCER research—it seems an unlikely mix on first encounter, but there's an intriguing story here that illustrates how curiosity-driven experimental science works to promote human wellbeing. Many of us will know about the link between papillomavirus infection and cervical cancer in women, a discovery that led ultimately to the development of a protective vaccine and to Harald zur Hausen being awarded the Nobel Prize in 2008. We may also be aware that the effectiveness of many cancer treatments has improved greatly over the past decades, though the overall incidence of most human tumours has gone up because we are living longer due to a whole spectrum of research-based advances in treating cardiovascular and other diseases. What we may not know, however, is that some forms of avian cancer have been known to be infectious for a century or more. Why is that of interest? Well, it turns out that, over the decades, studies of these virus-induced bird tumours have massively advanced our basic understanding of carcinogenesis (the process of cancer development) and have pointed the way to improvements in evidence-based cancer therapy.

Though I've lived through much of this as a scientific spectator working in a different research field, it was only when I started talking about these avian tumour models to cancer molecular biologists like my St Jude colleague Charles (Chuck) Sherr that I realised how important the bird viruses have been in illuminating the genetic events that result in cells escaping the normal mechanisms of growth control, the process that is basic to carcinogenesis.

But don't get the idea for a moment that human beings catch cancer from eating chickens, or from consorting with pet hens, budgerigars, parrots, ostriches or any other variety of our feathered friends. Zoo leopards can develop influenza when they're fed the carcasses of infected birds, but that link does not hold up for avian tumour viruses and cancer. The fact that some tumours do have an infectious cause is also true for cats, dogs, pigs, parrots, cows, frogs and just about any species you can think of. And while these cancer viruses tend to be very species-specific, the types of molecular events within cells that lead to oncogenic (cancer-causing) transformation have many common features.

Through the nineteenth century, scientists like Louis Pasteur, Robert Koch, Ronald Ross and a host of others worked out the role of germs in infection and dispelled forever the idea that diseases like malaria and cholera result from the spontaneous generation of 'animacules' or from exposure to 'miasmas' emanating from swamps and bogs. As we saw in Chapter 10, experiments with birds played a significant part in this process. Cancer, though, was considered to be non-infectious and in a completely different category. That's still true today for most, but not all, human tumours.

Oncologists generally have little interest in infectious disease, except from the aspect of the clinical problems that can occur when the cytotoxic drug and radiation therapies used to kill the rapidly dividing tumour cells also destroy, for a time, the patient's immune system, which normally keeps potential invaders (viruses, bacteria, fungi) at bay.

Cancer tends to be a disease of ageing, with the underlying molecular changes often emerging progressively as a consequence of accumulated mutational 'errors' in the cell. Geriatric solid tumours are also seen in pet cats, dogs and parrots, but chickens tend not to lead long lives. What's more, we cram domestic poultry together under conditions where any infection will transmit very readily. Tumour viruses, then, can be a big problem for commercial chicken production.

* * *

Way back in 1908, two Danish veterinarians, Vilhelm Ellerman and Oluf Bang, reported that they could transfer a form of avian leukosis through multiple passages in chickens. That didn't cause much of a stir at the time because it wasn't until the 1930s that leukosis, or leukemia (the unrestricted growth of white blood cells), was recognised as being a form of cancer. Aware of Ellerman and Bang's findings, the American Peyton Rous, working at the Rockefeller Institute, showed in 1911 that he could transmit a solid tumour of chickens (a sarcoma) by injecting healthy birds with an extract made from the ground-up cancer cells. Rous used the ordinary filter paper that you would find in any high school chemistry lab to exclude the possibility that he was actually doing a live tumour cell transplant.

The finding was dismissed by the oncologists of the time as a bird oddity that had no real relevance to human disease, and even Rous himself wasn't all that convinced that he had found something important. Then World War I came along and Rous focused much of his research effort over the next 15 or more years on physiological studies of the blood and liver. (Among his contributions, he helped to establish the first blood bank near the front line in Belgium.) After the conflict was over, there were (as always) plenty of other science problems to pursue, and he didn't enlarge on his initial bird and tumour findings.

The situation changed dramatically when his younger colleague Dick Shope (the Rockefeller Institute scientist we met in Chapter 7) established in 1931 that a skin cancer of rabbits could be transmitted by a filterable extract. That immediately excited Rous. The finding looked good: the rabbit study had been done using the more advanced technology of Berkefeld filter candles, and this discovery of what was soon called Shope fibroma virus showed that infection-induced cancers can also be found in mammals and are not just some odd feature of avian species. Then, in 1933, Shope further established that rabbit papillomas (warts) are caused by another quite different virus.

Shope papillomavirus is the prototype for similar pathogens that cause warts and tumours in a variety of vertebrates, including cattle and humans. The psittacines (parrots), in particular, can develop obstructive, infectious lumps in the throat and vent (cloacal) regions, but papillomaviruses are not generally a big problem in birds. Because papillomaviruses replicate only in keratinising epithelia (effectively dying skin cells), it took years to learn how to grow them in tissue culture, so their importance for understanding the fundamental biology of cancer has only been recognised quite recently.

Back at the Rockefeller Institute in the lead-up to World War II, Dick Shope was very busy studying more conventional infections (like influenza), and being a generous spirit, he happily handed over the analysis of his cancer-causing viruses to his friend Peyton Rous. As Rous notes in his 1966 Nobel lecture, Shope did not return to his rabbit tumours in any significant sense until after Rous finally closed his laboratory. At 87, Rouse was the oldest

Nobel Laureate on record, having waited 55 years for the initial chicken sarcoma finding to be recognised by the Nobel committee. As is often the case, Rous continued throughout his career to contribute significantly to the study of transmissible tumours, but he made no other major breakthrough. The next big cancer discoveries that depended on the use of viruses were to come later, and it was the Rous sarcoma virus (RSV), which he'd isolated way back in chickens, that provided the key.

With the passage of time and a lot of effort, it was firmly established that viruses with some of the general characteristics of RSV could cause tumours in a variety of mammalian species. Working at 'Mega Mouse', the Jackson Laboratories in Bar Harbor, Maine, Joseph Bittner found in 1938 that a mammary tumour of mice could be transmitted to offspring via the milk. Then, beginning in 1953 with the work of Ludwig Gross, a whole spectrum of murine leukemia viruses were found that confirmed Ellerman and Bang's original observation in chickens. Named after their discoverers, the Gross, Moloney, Abelson, Friend and Rauscher viruses (like Rous sarcoma) captured the attention of many investigators, although no one yet understood how they caused disease.

By the 1960s, with all this accumulating evidence, people were starting to think that viruses might indeed be a common cause of human cancer. That idea was very fashionable in 1971, when US President Richard Nixon declared the National Cancer Institute's 'war on cancer' one of his most important legacies. Playing a prominent role in the initial National Cancer Institute strategy was the 'Special Virus Cancer Program' (headed by George Todaro and Robert Huebner), where the young Chuck Sherr did his first substantial experiments in the field.

A great deal had, of course, happened in the 60 years between 1911, when Peyton Rous first showed that his humble chicken tumour was transferrable with cell-free extracts, and 1971, when Richard Nixon signed the war on cancer into existence. The single most important advance in biology was the 1953 discovery of the molecular nature of inheritance by Jim Watson and Francis Crick, who were then working at Cambridge University's Cavendish Laboratories. Inspired by Rosalind Franklin's X-ray crystallographic pictures (see Chapter 9), which Jim sighted (unbeknown to Rosalind) during the course of a visit to Maurice Wilkins at Kings College, London, Watson and Crick built the iconic physical model that shows how binary pairings of the deoxyribose nucleic acids (DNA)—adenine–thymine (AT) and guanosine–cytosine (GC)—can assemble as a double helix. This provided an immediate explanation for inheritance. To put it very simply, the information that passes from one generation to another is specified in the DNA triplet code of our genes. The genes are assembled on larger, paired structures called chromosomes, which can be seen down a light microscope after appropriate staining and differ in number from species to species: humans have 46, chimpanzees 48, kangaroos 12, hedgehogs 88, chickens 78, turkeys 82 and so on. Separation of the two DNA helix strands established how it is that we, and the birds, inherit about one-half, or haplotype, of each chromosome from our mother and the other from our father. I say 'about' because, unlike the sperm, the ovum also contains mitochondrial DNA that always descends in the female line. The same is true, of course, for hens and roosters. Mitochondrial DNA is a powerful tool for tracking hereditary pathways in both humans and birds.

By the time that Watson and Crick were awarded the 1961 Nobel Prize, poor Rosalind Franklin was dead of ovarian cancer and they shared the stage in Stockholm with her somewhat alienated King's College colleague, Maurice Wilkins. Events moved very quickly, and a central dogma soon emerged that information is copied (transcribed) from the genetic material of the chromosome (the DNA) to produce a ribose nucleic acid (RNA) 'message'. The sequence of 'messenger' RNA nucleic acid triplets in turn serves as a template for the linear assembly (translation) of amino acids, the building blocks of the proteins that are the key structural units of biology. What also happened over these years is that the mouse and avian tumour viruses had—like influenza and yellow fever virus—been shown to carry their primary genetic material as RNA rather than DNA, a conclusion that led to their being classified as riboviruses.

Geneticists had at this point come to what we now recognise as an over-simplistic conclusion: that errors in nuclear DNA (not RNA) call all the shots when it comes to the aberrant patterns of cell differentiation and growth control that lead ultimately to cancer. If this was the case, though, it was hard to understand how RNA viruses could induce tumours. That was less of a problem for the DNA viruses, like Shope fibroma and Shope papilloma, but as is sometimes the case in research, it turned out that tackling the more conceptually difficult RNA tumour virus question opened up the whole field in totally unexpected ways.

This is where Peyton Rous's little chicken virus came into its own, and for a rather simple reason. The key point about RSV was that it could be grown to very high titres in living chick embryos and in cultured chick embryo 'fibroblasts', giving the scientists a lot of 'stuff' to work with. The fibroblast technique produces a neat, contiguous sheet—or monolayer—of embryo cells. When the monolayer is exposed to even a single particle of 'lytic' virus (such as yellow fever or influenza) and is immediately covered in a semi-soft agar overlay, scientists are able to observe replication of the virus through direct cell-to-cell contact. As the virus replicates, the infected embryo cells die, creating holes (or 'plaques') in the monolayer.

Research along these lines was ongoing at Caltech, Los Angeles, in the virus laboratory headed by the expatriate Italian scientist Renato Dulbecco. Dulbecco's close colleague, the veterinarian Harry Rubin, introduced a young graduate student, Howard Temin, to the RSV model. Working together, Temin and Rubin had the insight to plate out the chick embryo cells at concentrations that were relatively sparse and did not form a monolayer.

This allowed them to see that the virus caused some of the embryo cells to proliferate, round up, and in effect form little RSV-associated tumours. Instead of punching holes in the monolayer, RSV had caused the production of 'lumps'. That was a real break-through: they'd managed to move this virus-induced cancer from the enormous complexity of the living chicken to a simple, readily investigated cell culture system.

Temin went on to conduct a key experiment, adding the DNA inhibitor actinomycin D to his RSV-induced 'cancer' tissue cultures, with surprising results. Though actinomycin D doesn't block RNA, it completely inhibited tumour production in the RSV tissue cultures. Shortly after, Temin proposed at a scientific conference that the infecting RSV somehow acts as a template for the synthesis of a DNA 'provirus', which is then acquired as an integral part of the host cell's genetic make-up. At this meeting, and for the next six years, his highly heterodox proviral DNA hypothesis was mostly ignored. But one person who didn't think Temin was crazy was the young David Baltimore, who was then working with one of the murine leukemia viruses at the Salk Institute in La Jolla, California.

In 1970, Howard Temin and his Japanese postdoctoral fellow Satoshi Mizutani showed, in concurrent publications with David Baltimore, that these RNA tumour viruses carry a protein (reverse transcriptase) that functions to allow elements of the viral RNA to be copied back into host DNA. This began our under-standing of how the DNA-tumour rule might be satisfied for the oncogenic RNA viruses, now called the 'retroviruses'. Temin and Baltimore didn't have to wait as long as Peyton Rous for the Nobel Prize, which they shared with Renato Dulbecco in 1975. Later, when another (very deadly) RNA virus hit in the 1980s, much of the biology of HIV infection was explained by the fact that it also carries a reverse transcriptase. Apart from that, just about every biomedical research and diagnostic laboratory in the world uses reverse transcriptase to convert messenger (m) RNA to a complementary (c) DNA, which is then expanded by Kary Mullis's PCR reaction (see Chapter 7).

But while Temin and Baltimore had found the reverse transcriptase enzyme, they hadn't sorted out how the proviral DNA suggested by Temin caused rapid sarcoma development. That fell to Mike Bishop, Harold Varmus and their team at the University of California, San Francisco.

By the time Bishop and Varmus came to the problem, others had put together a story that a gene (called *v-src*) within RSV was in some way causing the transformation to cancerous growth, probably by directing the synthesis of a novel 'oncoprotein'. Drawing on the previous work of Steve Martin, Peter Duesberg, Peter Vogt, Hideo Hanafusa and colleagues, and on a crucial experiment conducted by a young French postdoctoral fellow by the name of Dominique Stehelin, Bishop and Varmus were able to identify the normal-cell equivalent of *v-src*, which became known as *c-src*. This normal *c-src* gene was present in avian species as distant as the ratites (like the ostrich and the emu). Discussions with a Berkeley colleague, Allan Wilson, led them to deduce that *c-src* had been conserved through the 100 million years or so of evolution that separate the domestic chicken from the Australian emu. Some time in the past, RSV picked up the *c-src* gene from normal bird cells and it became *v-src*.

Bishop and Varmus, with the help of Vogt and Stehelin, had found the first of the 'oncogenes', growth-promoting genes that, when mutated or inappropriately expressed at abnormally high levels, turn a normal cell into a cancer cell. Subsequent discoveries defined many additional retroviral oncogenes, likewise christened for the (occasionally exotic) types of tumours that they induce: *v-ras*, *v-mos* and *v-fes* (producing rat, mouse and feline sarcomas, respectively), *v-myc* (myelocytomatosis), *v-myb* (myeloblastosis), *v-erb* (erythroblastosis) and so forth. (For 'omatosis' and 'blastosis', read blood cancer.) In each case, proviral integration had allowed the virus to acquire a normal cellular gene and, with it, the capacity to rapidly transform any host cells infected by newly produced viral particles. As Mike Bishop would later say, 'the seeds of cancer are within us'.

It turns out that the *src* gene and some of its later-defined relatives encode a family of growth-promoting enzymes called tyrosine kinases (TKs), and that the mutated *v-src* version is more potent than its normal *c-src* counterpart. The realisation that aberrant TK expression promotes cancer enabled the development of the TK-inhibitor imatinib (marketed as Glivec), which is a potent chemotherapeutic agent for treating chronic myelogenous leukemia, gastrointestinal stromal tumours and some other forms of human cancer. In the future, we expect to have many more therapeutics like imatinib, which (as with the Relenza antiviral discussed in Chapter 9) is one of the first 'designer drugs' developed from a prior knowledge of the chemical structure and function of a protein that is found normally in nature.

While RSV has brought us a great distance in understanding why some cells escape normal growth control and cause tumours, it turns out that the related retroviruses are not a major cause of human cancer. Even so, studies of Peyton Rous's little chicken cancer virus have led us to a number of important breakthroughs: the understanding that some forms of cancer are contagious; the discovery of the reverse transcriptase that first showed how genetic information encoded in RNA can be copied back into the DNA of the genome and later opened the way for Kary Mullis's critically important PCR technique; the identification of the cellular oncogenes that play a central role in the formation of tumours; and finally, the possible development of a whole new spectrum of therapeutic agents for treating cancer—not a bad outcome for a discovery that, way back in 1911, didn't seem to have much practical value.

14

Blue bloods and chicken bugs

TRADITIONALLY, ANY NOTION OF good breeding and inheritance was synonymous with the idea of blood lines. That was before the rediscovery of Gregor Mendel's sweet pea breeding experiments (obscurely published in 1866) led to the founding in 1901 of the science of genetics, and way, way before Watson and Crick had explained the nature and function of DNA. Even so, many still equate blood and heredity, particularly when it comes to thoroughbred horses, dog breeds and aristocrats. You sometimes hear it said that 'blood will tell', or 'blood will out', which has nothing to do with bleeding, or with murder for that matter.

Over the ages, the principal of hereditary monarchy has depended on the assumption that 'royal blood' confers superiority and consequent privilege. But being a blue blood carries no guarantee of genetic superiority in humans, at least so far as disease is concerned. (The idea that royalty and members of aristocratic European families were 'blue blooded' probably derives from the fact that, not having to work in the fields, their blue, venous blood was visible through pale, white skin.) The intermittent madness of George III (1738–1820) is thought to have been a consequence of variegate porphyria, a condition that probably came down to him via his ancestor James I of England, also known as James VI of Scotland (1566–1625). And it is certain that Queen Victoria passed on the X-linked gene for haemophilia. Being on the female X chromosome, this disease is only manifest in any XY male who is unfortunate enough to get the abnormal X, while the XX female is protected by the alternative, 'silencing', normal X.

The sex chromosome equation is the other way around in birds, with the females being heterogametic (ZW), while the males are ZZ. That means that any Z-linked genetic abnormality shows up in the hen rather than the cock. Haemophiliac birds would not last long, of course, either in nature or in a human-directed breeding program, so the persistence of a Z-linked mutation causing such a severe health defect would simply manifest as the early death of a few females. Even so, sex chromosome linked genetic effects may not be all bad. The Z-linked lutino characteristic in cockatiels is determined by a disabling mutation in a grey pigment (melanin) gene, but this doesn't manifest as albinism, because (unlike us) the birds also have yellow pigment genes (psittacins). The mutant melanin gene is recessive, meaning that in a male that has only one copy, it will be 'silenced' by the normal Z. Any female bird bred from a lutino male will thus be yellow, because the cock must have two abnormal Zs, while that will not necessarily be true for the female offspring of a lutino hen (and never the case for the males) when the parent cock has two normal Zs.

Inheritance may thus have good and bad consequences, but can 'blue bloods' and favourable 'blood lines' really be characterised by anything that can be detected in blood? Thinking about measurement and difference in this context immediately brings the blood groups to mind. Working in Vienna, Karl Landsteiner discovered in 1900 that agglutinins (antibodies) in serum could be used to define the now familiar human ABO types of the erythrocytes, or red blood cells (RBCs). By 1907, it had been shown that ABO matching is predictive of successful blood transfusion—a finding that was central to Peyton Rous's efforts to establish a blood bank to save soldiers wounded in the trenches of World War I.

Generally weak correlations between human ABO types and disease susceptibility have been described, but when it comes to using molecules expressed on red blood cells (RBCs) as genetic markers for superior resistance to infections, it's among the chickens that we find the true 'blue bloods.' Furthermore, an understanding of why this is so provides important insights for understanding the genetics of susceptibility and recovery patterns in human disease.

By the 1930s, researchers were beginning to sort out the blood groups of birds. There was no interest or practical benefit in bird blood transfusion, of course, but following the 'climb Mount Everest' principle—because it's there and because it's possible—scientists will always ask such questions. And, after all, blood is readily obtained and blood groups are easy to measure. Finding a correlation between blood type and some highly desirable characteristic would also provide a great tool for domestic chicken breeders. And that's exactly what happened. But before we begin that story, a few definitions might assist those who have not been trained in biology.

* * *

As explained in Chapter 13—and put very simply—information that passes from one generation to the next is specified in the DNA triplet code of our genes, and these genes are assembled on chromosomes. We inherit one half, or haplotype, of each chromosome from our mother, and one from our father. The particular site where a gene is located on a chromosome is called the locus, while the various versions of that gene that might be found in different individuals are called alleles. The alleles we get from both parents may often be the same, reflecting that the structures (proteins, for example) that they encode can't readily be varied, so that they are conserved throughout the species and, sometimes, through evolution. On the other hand, the genes (alleles) at some loci are highly variable, or polymorphic.

Such polymorphisms are very characteristic of the immune system's antibody (immunoglobulin, Ig) and T cell receptor (TCR) genes, which encode the millions of different cell-bound TCRs and secreted Ig molecules. That incredible diversity of T cell (TCR) and B cell (Ig) 'repertoires' is essential if the immunological 'orchestra' is to find the 'notes' for a 'score' that satisfies the needs of a multifaceted 'audience', the organ systems that can be invaded by an infinite spectrum of viruses and other bad bugs (the 'muzak' of the microbiological world, if you like). Both in birds and mammals, TCR diversity is set up via molecular mechanisms that operate during development in the thymus, the organ in the neck region that shrinks as we get older. Immunoglobulins, like the virus-specific neutralising antibodies and blood group agglutinins that circulate in the serum fraction of the blood, are secreted by specialised, long-lived plasma cell 'factories' that can hang around forever in various anatomical sites throughout the body. Using strategies that are in some senses comparable to those operating in the thymus, the precursors of the plasmas cells (B lymphocytes, or B cells) develop first in the mammalian bone marrow, or in a specialised avian organ (the bursa of Fabricius) located near the cloaca.

Over 150 million or more years of evolutionary divergence, the avian and the mammalian 'adaptive' immune systems have come to be organised somewhat differently. The ultimate functional outcomes, however, are identical, with both systems doing the same job of countering the deleterious consequences of invasion by different pathogens. Reflecting evolutionary changes that were in place some 400 million years back, all vertebrates, from the bony fishes and beyond, share the characteristic of specific immune T cell and B cell (plasma cell) memory, ensuring resistance to reinfection and making vaccination possible.

As in humans, several different blood group systems have been defined in birds. While alleles encoded at the ABO locus on chromosome 9 define the blood groups that are most prominent in us, for chickens it's the B locus on chromosome 16. (The use of 'B' to describe this locus doesn't imply any relationship to the 'B cells' we met earlier in this chapter, just that the scientists who found them had to call them something.)

In fact, apart from the different chromosomal location (and that numbering is quite arbitrary) the ABO and B blood group antigens (structures that bind antibody) are fundamentally distinct in character. The allelic variation that defines the ABO system is expressed in oligosaccharides (sugars) that are attached to proteins in the outer, plasma membrane of the RBCs. On the other hand, the chicken B locus antigens expressed on RBCs are cell-surface glycoproteins. The 'glyco' refers to the bit of carbohydrate (a sugar, like sialic acid) that is attached to the protein, but the essential differences are in the proteins themselves.

The early chicken geneticists tended to work closely with commercial breeders, so the rapidly accumulating evidence that different B locus alleles are predictive of increased resistance to a number of infections was both intellectually fascinating and an important practical breakthrough. The list of infectious agents with B allele–associated susceptibility profiles includes bacteria (*Salmonella* and *Staphylococcus* species), round worms, avian infectious anemia virus, infectious bursal disease virus, Rous sarcoma virus and Marek's disease virus, which causes another transmissible cancer that can cause major problems for the poultry industry. Furthermore, particular chicken B types are associated with improved immunity following different vaccinations. Nothing so substantial in the sense of predicting host response profiles has ever been seen for the human ABO system.

It turns out the main thing that the chicken B locus and the ABO blood groups have in common is that researchers first found them on the surface of RBCs. Otherwise, we would tend not to link them in our thinking. The chicken B locus proteins are not, in fact, specifically blood groups, but are the same transplantation, or major histocompatibility complex (MHC), molecules that can be expressed on almost every cell throughout the body.

Largely as a consequence of the discovery we made back in the mid 1970s, people had been looking for years to find strong links between particular MHC genes and disease susceptibility profiles. Analysing a problem in viral immunity, Rolf Zinkernagel and I found by chance that immune T lymphocytes recognise virally modified MHC glycoproteins. If you want to know how that works in the molecular sense, the story is told in our Nobel Lectures (available at the Nobel e-museum website) and in my book *The Beginner's Guide to Winning the Nobel Prize*. This infectious disease–MHC association is seen easily in chickens as they have only the one relevant B locus. Humans and mice each have three different MHC loci: HLA-A, B, C and H2K, L, D respectively. As there are usually two different alleles (genes) at each locus, it's easy to see why a disease association is more apparent when there are only two possibilities to choose from (as is the case with chickens) rather than six (with humans).

The chicken B locus story has thus shown us that there are genuine genetic 'blue bloods' when it comes to birds and immunity. Using either conventional breeding or sophisticated molecular biology approaches (transgenic technology), it is possible to introduce resistance genes and to disseminate them throughout a species at risk. In January 2011, for example, it was reported that scientists had successfully produced transgenic chickens that are much less susceptible to avian influenza. Thinking about such processes and breakthroughs in the context of humans raises major ethical problems that are way beyond the scope of this book. There are also ethical questions when it comes to modifying domestic animal species using genetic engineering strategies, though once the question of food safety is addressed, these are much less profound. At least, that's true so long as we're not trying to breed more 'productive' monsters (chickens with four legs!) and so long as we're thinking purely in terms of infection and immunity.

15

Killing the vultures

VULTURES ARE NOT PARTICULARLY lovable birds, and they don't generally receive much favourable press. The cartoon image that comes to mind is of the guy crawling in the desert, tongue hanging out, no water in sight, while a couple of bald-headed, gloating vultures perch close by waiting for his final collapse and loss of consciousness.

But the fact of the matter is that vultures are, like the much more highly regarded bald eagles, essential scavengers that clean our landscapes. In Africa they share that role with some mammals, particularly the hyenas, though to be called a hyena is no more of a compliment than to be described as a vulture. The African vultures, and no doubt the hyenas, are thought to have coevolved with ungulates like the buffalo, wildebeest, springbok, giraffe and zebra. As the human population continues to expand, these native herbivores progressively lose habitat, and more and more are killed to provide 'bush meat'. Some species (particularly wildebeest and buffalo) are now farmed, but with directed harvesting and management, many fewer will fall to lie in the field. The inevitable consequence is a decrease in vulture counts everywhere but in the well-monitored wildlife parks where the natural cycle of life and death is rigorously maintained. In addition, being big birds like the bald eagles, vultures tend to fly into high-tension wires and other obstacles that we place around the landscape.

The African bearded vulture is classified as endangered, while the more common white-backed vulture has been put in the near-threatened category by the International Union for the Conservation of Nature (IUCN). Overall, though, the situation of vultures is much more dire in India, Nepal and Pakistan. The IUCN's red-list category identifies the three species of Indian vulture as critically endangered, a situation that has developed with enormous speed over the past 20 years or so. Apart from threatening the birds themselves, this is also an extremely serious situation for the extensive rural populations of these countries. There are no hyenas on what is generally known as the Indian sub-continent, but there are many cattle, which are, partly for religious reasons, often left to die naturally. The vultures have traditionally provided a completely free and very efficient sanitation service.

As recently as 1992, there were vultures everywhere in the rural areas of the subcontinent. Then, before that decade was out, the numbers fell very rapidly. Both concerned individuals and local authorities started to look closely, and a number of US groups, like the Peregrine Fund based in Boise, Idaho, and the Washington State University Veterinary School at Pullman, quickly became involved. Dead birds were collected, preferably close to the time of their demise—that part of the world tends to be pretty hot on the whole, and as I learnt in my early days as a veterinary scientist, the fresher the better when it comes to doing a detailed post-mortem.

The investigating pathologists quickly established that once the usual causes of avian exit were excluded, otherwise healthy vultures were dying acutely of visceral gout. Gout? For most of us that evokes the image of some testy, bibulous, red-nosed old duffer with his painful, swathed foot and ankle parked on a cushioned footstool. Vultures aren't likely to imbibe copious quantities of port and red table wine, so what could be happening here? Well, human gout is caused by the accumulation of needle-like uric acid crystals in joint tissues and blood capillaries. The ultimate break-down product of the purines found in many foods, uric acid is normally excreted via our kidneys. Conditions associated with progressively compromised kidney function, like hypertension and diabetes, predispose to gout. If there's no underlying genetic or other medical cause, gout may be more a consequence of a long-term over-indulgent, low-exercise lifestyle characterised by the excessive consumption of red meat and salt. Drinking a lot of good (or bad) wine can go with that, but alcohol may not be a primary cause. And even abstemious vegetarians aren't totally protected, as an excess intake of fructose is also a risk factor.

Birds and humans share the characteristic of not having a functioning urate oxidase (uricase), the enzyme that breaks down uric acid. It's true that vultures do go for the red meat bit, but they also get plenty of exercise and they've consumed that diet through evolutionary time without having a notable gout problem. Also, the dead Indian birds had visceral gout, meaning the uric acid crystals were so widely distributed that the organs looked white when the pathologist opened the body cavity. Of course, all birds excrete a concentrated white paste of uric acid rather than the dilute solution of urea (and some uric acid) that is characteris-tically found in mammalian urine. The presence of uric acid all over the place immediately suggested that some form of kidney damage must have been responsible for the vulture wipeout.

Given the acuteness and wide territorial distribution of the problem, the first thought was that the vulture deaths might have been caused by a novel infection. Working with investigators in Pakistan and India, the Washington State microbiologist Lindsay Oaks and Scots veterinarian Martin Gilbert took on the task of exploring this possibility, but they could not identify a causative virus, fungus or bacterium. Samples were also sent to the ultra–high security Australian Animal Health Laboratory (AAHL) in Geelong, about an hour's drive from Melbourne. The AAHL virologists really know their stuff, and they did find a new herpes virus, but though there were concerns that this could prove a problem for captive vulture breeding programs, it was clearly not inducing the visceral gout problem.

Failing to identify an infectious cause, Oaks, Gilbert and Peregrine Fund biologist Munir Virani reasoned that dead domestic livestock were a major food source for the affected vultures. Could the cause be something that was being given to the cattle? They checked out the veterinary drugs that were currently in widespread use, focusing on those that were relatively new to the market and that might be expected to cause some degree of renal toxicity. That led them to the non-steroidal anti-inflammatory drug (NSAID) diclofenac (Voltaren). Oaks tested his samples and found that all those from birds with visceral gout contained diclofenac at some level, while the 'control' tissues from an equivalent number of gout-free vultures were uniformly negative. Their report appeared in the leading science journal *Nature* in 2004. (Even the world's top scientists break out the champagne if they get a paper published in *Nature*, and what editor could resist a great vulture detective story?)

Where was the drug coming from? Diclofenac is out of patent, is very cheap, and was being made by a number of Indian manufacturers. Though cattle enjoy a particular status in Hindu culture, they are nevertheless put to productive use. Loads of the drug were being given as a broad-spectrum treatment for general inflammatory conditions and malaise. Draft oxen (castrated cattle) are still very important to the rural economy, and treatment with diclofenac meant that those with sore joints could soon go back to work. It was also used to restore milking cows to production. And though they are not to be killed, it is acceptable to provide some treatment for cattle that are roaming free and obviously suffering. Many such animals are grossly compromised, not least because of the ingestion of plastic bags—cows are not picky eaters.

There's so much happening in science, and most of us are far too busy. Having missed the *Nature* paper by Oaks and his colleagues and the excellent overviews written by Susan McGrath and others, I was, in fact, completely unaware of the vulture story until I made a 2009 visit to the University of Pretoria's Veterinary School in South Africa. I was there at the invitation of the Dean, Gerry Swan, to give the Sir Arnold Theiler Memorial Lecture. This was an offer that could not be refused. Arnold Theiler—the father of Max Theiler, the yellow fever vaccine pioneer whom we met in Chapter 4—is a genuine hero of infectious disease research.

During my visit, I got to talk science at length with Gerry and with a number of the senior faculty members. Hearing about what's happening with animal diseases, particularly those that don't transmit to humans, is often fresh and interesting, as I'm pretty much embedded in a basic biomedical research culture. My discussion with Gerry and his close colleague and protégé Vinny Naidoo was my introduction to the Indian vulture story.

Gerry is a pharmacologist who returned to academia after spending some years working for what was then Merck & Co. His expertise in drug research, together with the certainty that the African white-backed and griffon vultures were not experiencing the visceral gout problem, made him an ideal person to look further at the diclofenac issue. Responding to a request from Mark Anderson of the Endangered Wildlife Trust's Vulture Study Group, Gerry and his research group soon reproduced the characteristic visceral gout pathology by dosing African vultures with diclofenac.

Using the African white-backed vultures, Swan, Naidoo and their colleagues were then able to show very quickly that an alternative and equally inexpensive anti-inflammatory, meloxicam, could be used in cattle without causing any problem for the big birds. Even with this breakthrough, though, and with the Indian and Pakistani governments imposing a total ban, it's been very hard to get the message out and to remove all the diclofenac from the farms and veterinary supply houses. In addition, another NSAID, ketoprofen, has come into the picture and is just as toxic as diclofenac.

These are good drugs so far as cattle are concerned, but when any therapeutic is given as a broad-spectrum, symptomatic treatment to large numbers of animals in the field, it is inevitable that a few will die from some underlying, and often undiagnosed, cause. Hundreds of vultures can be observed to feed on a single dead cow. According to the mathematical ecologists, the massive, rapid decline in the Indian vulture populations could be accounted for by as few as one in every 760 cattle carcasses having substantial levels of diclofenac.

How does the drug kill? Like many toxins and therapeutics, it concentrates in the liver, which is like pâté de foie gras to a hungry vulture. The mode of action is still controversial, and I won't confuse you and possibly myself by discussing the likely mechanisms in any depth. Suffice it to say that there is damage to (necrosis of) the sensitive, proximal, convoluted kidney tubules, which could be a consequence of diminished blood flow and/or the effect of those terrible free radicals we hear so much about in advertising blurbs for non-prescription medications. Whatever the final cause, the resultant failure to clear uric acid via the kidney leads to its rapid, lethal accumulation in tissues. This in turn causes excessive potassium build-up in the blood and terminal organ failure.

The principal consequence of the die-off in the vulture populations is, of course, the loss of a free and efficient environmental sanitary service. The net result is greater food availability for other carrion-eaters. Wild dogs, which are both abundant in India and the main carriers of rabies, increased dramatically in numbers. As a consequence the risk of dog-bite also went up for rural workers, with a 2008 estimate suggesting that the loss of the vultures led to 50 000 excess human deaths in India alone. Rabies is a horrible way to die. There is the possibility of post-exposure treatment, but cost, limited medical access and the lack of education that causes poor farmers to opt for 'traditional' remedies means that being bitten by a rabid dog is too often a death sentence. Aggressive programs to cull or sterilise dogs that are not properly controlled are leading to substantial mitigation of this problem.

But the loss of the vultures also has cultural repercussions. The Parsi have long followed a very reasonable practice of putting their dismembered dead out on stone 'towers of silence' to be stripped by vultures. No vultures means no corpse removal. Also, as India becomes increasingly prosperous and more people take the spectrum of pills that are already ingested by the elderly in Western societies, one wonders what other drugs may prove toxic for avian carrion-eaters. Though the Parsi practice has the great advantages of not taking up the available land and causing no greenhouse gas emissions, maybe it's time to consider another approach.

Other birds, like the corvids, are also susceptible to the toxic effects of the NSAIDS, and that characteristic visceral gout must now be on the watch list for veterinary and wildlife pathologists everywhere. One way or another, we are putting enormous amounts of potentially toxic materials into the natural environment while making no effort to assess the possible effects on wildlife and other species. NSAIDS like ibuprofen are, for example, also toxic for dogs. Dogs and vultures are very visible, but what of the myriad other animals that inhabit our world?

16

Heavy metal

IT'S A MATTER OF personal taste whether you find heavy metal music to be toxic, but there can be no doubt about the real thing. Metals can be lethal, but they aren't all bad—and maybe that's also true of heavy metal music, if you belong to the right generation, or if it's played outdoors and you're far enough away from the band. Similarly, when it comes to metals and vertebrate biology, levels, context and location are everything.

Some heavy metals have no normal, physiological part to play in bird or beast. That lists includes lead, mercury and cadmium. In the past, lead compounds, such as white lead and lead acetate, have been used in cosmetics, as a contraceptive and for human hair colouring. Arsenic is another heavy metal that, in various formulations, was widely used in traditional medicine and in the early days of pharmacology. The 'father of immunology', Paul Ehrlich (who won the Nobel Prize in 1908), developed the long-supplanted drug arsphenamine (Salvarsan) for the treatment of syphilis and trypanosomiasis, while other arsenic compounds were commonly used in face powders, as appetite stimulants or as a foaming agent in beer. Both birds and people die from ingesting arsenic-contaminated groundwater, a particular problem in India and Bangladesh, while spraying arsenates on crops for pest control also puts birds at risk. The World War I poison gas Lewisite is an organoarsenic compound that causes severe lung damage. The antidote, British anti-Lewisite (BAL), is a chelating agent that binds to a spectrum of heavy metals (including lead and arsenic) so that the complexes are excreted into the bile, then out through the gastrointestinal tract.

Some metals, on the other hand, are essential for normal body function. That's certainly the case for iron, copper, zinc, cobalt, sodium, potassium, manganese, magnesium and molybdenum. But all are normally complexed with other molecules to form physiological salts (like sodium chloride), or they act as cofactors with, for example, proteins that are essential for oxygen (O_2) transport (such as iron in haemoglobin) and use (such as copper in cytochrome C oxidase). Zinc is ubiquitous and is essential for many of the enzymes, transcription factors, signal transduction molecules and so forth that are vital to the function of all cells in the body. The superoxide dismutases, which operate naturally to neutralise dangerous free radicals (like superoxide O_2-), variously contain copper, zinc, manganese or iron.

Molecules like transferrin act to maintain adequate levels of iron and to prevent the development of anemia. Excess iron absorption from the gut in the heritable condition haemochromatosis leads to physiological overload and, eventually, to complications like liver cirrhosis and cancer. Part of the treatment is regular bleeding to remove some of the iron. A haemochromatosis-like disease has been described in wild European starlings and in some captive tropical birds like mynas and toucans, though the cause is not known. In humans, a genetic defect that leads to decreased copper excretion also leads to the severe liver pathology and consequent brain damage characteristic of Wilson's disease. Treatments include the administration of chelating agents like BAL or penicillamine to remove the excess copper. If that fails, the ultimate, drastic recourse is a liver transplant.

Heavy metals can be very toxic when ingested in free form or as a soluble salt. In the case of caged birds, acidic foods and fluids can leach out copper, so copper vessels and pipes are best used with caution. Parrots should be discouraged from chewing on old pennies or on zinc-coated or nickel-plated bird wire. Painting birdcages with anti-rust primers containing zinc chromate is also a bad idea. Adding a little copper sulphate to poultry feed increases erythropoiesis (blood cell production), but feed mills have been known to get the concentration wrong with the result that the birds are poisoned. The classical laboratory experiments on copper toxicity were done years back by adding increasing amounts of copper salts to the water in fish tanks, then doing pathology on the brains of the unfortunate goldfish.

Birds in the wild function as a roaming, natural detection system for the presence of heavy metals. Rivers and estuaries can be contaminated with run-off containing methyl mercury produced in the course of industrial acetaldehyde production. This mercury derivative is concentrated by filter-feeders (molluscs and oysters) and then passed up the food chain, finally reaching maximum levels in the tissues of fish-eaters like cormorants, eagles and us. Methyl mercury is equally toxic for avian and mammalian brain cells. Found to be the cause of severe ataxia (incoordination), numbness, insanity, coma and human death in Minamata, Japan, as early as 1956, it took years before there was any effective preventive action, despite the fact that birds were dropping from the sky and deranged cats were committing 'suicide'. In Minamata, the source was local, so the levels in water were very high, but there are also potential consequences when such uncontrolled industrial run-off is diluted in the seeming vastness of the oceans. Tests on algae, shellfish, pelagic fish and feathers from tufted puffins and other seabirds sampled from the remote waters of Alaska's Aleutian Islands, for example, have shown increased levels of mercury, though not yet at levels sufficient to cause public health concern. The northern ocean currents are such that much of the massive debris resulting from the Japanese tsunami in 2011 has followed the same path.

* * *

We all know about lead poisoning. Some have even suggested that the use of lead pipes and lead cauldrons helped to bring about the fall of Imperial Rome. (Maybe the pipes weren't all that dangerous, as they would soon have become coated with an inner 'scale' of calcium carbonate.) Everyone is aware, though, of the dangers posed by flaking or cleaned-off lead paint on old houses. As with any ingested toxin, body mass matters, so birds and small animals, including children, are particularly at risk. Birds can and do act as sentinels when it comes to environmental lead poisoning. Indeed, the early, 'dirty' days of mining in the American West almost led to the extinction of the California condor, a species that is still threatened, but more of that later.

Another classic (and more recent) example of mining-related lead poisoning occurred in 2007 in the port town of Esperance in Western Australia. Dust from powdered lead carbonate concentrate settled in houses, on the ground and in rainwater tanks. (I grew up with exactly this type of pollution throughout my childhood in the outer suburbs of Brisbane, though the contamination in that case came from the local 'cement works' and was much less toxic than the Esperance variety. Even so, living year in year out with occasional palls of grey dust gave me a good understanding of how pervasive such contamination can be.) The dust in Esperance wasn't coming out of an unfiltered smoke stack but blew off the wharves, particularly during the process of loading. A comparatively simple smelting process that solidifies the material into ingots would have removed all risk. Another alternative would have been to bag the powdered concentrate and move it in sealed drums That's what happens now, but it wasn't the case then. In the final analysis, though, the people of Esperance were lucky. They got a warning, and both the medical authorities and the politicians were forced to act. What alerted them was the sudden death of some 4000 or more birds. The dust did not discriminate, killing wattlebirds, gulls, sparrows, in fact anything with feathers.

The only good thing about the Esperance event was that the nature of the problem was recognised before there was widespread poisoning, of children in particular, at levels sufficient to cause anemia and brain damage. Esperance householders still need to be careful with their use of rainwater and to treat household dust with respect. But, though this story is reasonably well known, it has not necessarily led to a generalised response. Australia has a federal system, so while the West Australian authorities were forced to deal with the situation, other states weren't. Recent media reports indicate that the citizens of Mount Isa, in Queensland, may also be at risk.

Maybe, then, if you live in a mining town or port and have some spare time, birdwatching would be a good hobby. It might also be sensible to find out which authority should be contacted if a sudden increase in bird deaths suggests a potentially toxic cause. Another approach, of course, is to watch out for the characteristic blue-grey gum line of lead poisoning in your kids, but buying a pair of good binoculars and spending some time out in the fresh air seems a better alternative in every respect. Also be aware that the symptoms of nickel poisoning include skin rash, dizziness, nausea and insomnia. Atmospheric nickel levels can increase as a consequence of a number of operations associated with the mining industry. The concentrations of environmental nickel were also found to be very high when lead contamination levels were analysed in Esperance.

* * *

We have taken the lead out of gasoline, and the mining companies are now generally subject to appropriate monitoring and regulation, at least in what we broadly regard as the developed nations, but there is still another major cause of lead poisoning in birds. And that relates to leisure activities, specifically the use of lead shot for hunting and lead sinkers for fishing. Both are progressively being banned throughout what we like to think of as the civilised world.

As a metal, lead is a bit unusual in the respect that it only remains bright and shiny for a very short time. Molten lead looks just like the silver mercury in an old thermometer, but it soon loses that lustre. It rapidly oxidises, producing its characteristic dull grey appearance. In addition, those who make their own bullets, shot or lead sinkers—a straightforward process, as the metal melts over a wood fire—will know just how rapidly that greying process occurs. In other situations, lead readily combines with acids to produce various salts. The point is that though lead itself does not dissolve in pure water, it is easily converted to various forms (acetates, nitrates) that are water soluble, and thus bioavailable if ingested or injected (as bird shot).

Besides cheapness, the other great advantage that lead has for sporting types is its density. That means, for example, that a 1-ounce load of BB shot will contain 50 pellets of lead, versus 70 pellets of steel or the like. An alternative shot formulation contains tungsten, iron nickel and tin (TINT). All of these are potentially toxic in isolation, but formed into a substantially non-degradable pellet, they are only a danger if the hunter actually finds his or her mark.

'What?' you may be asking. 'Surely lead pellets are only problematic if the bird is hit?' In fact, that's not the main issue. Most people are rotten shots anyway, and shotguns are usually 'choked' to give a greater or lesser spread. If a grouse, for instance, is hit with the whole of a 1-ounce load, there won't be much of the bird left for either display or eating. It's been calculated that for every 50 pellets that find their target, some 6000 fall to earth. Gravity sees to it that what goes up (or out) must come down. The consequence is that heavily used hunting grounds can have pellet concentrations ranging from 18 000 to 70 000 and, exceptionally, even 400 000 per hectare. One figure from the US Environmental Protection Agency suggests that some 70 000 tonnes of lead are deposited each year in outdoor shooting ranges. That's a lot of lead, and an enormous waste for a metal that is becoming scarcer and has other important uses, such as in batteries.

Many grazing and diving birds pick up small stones to aid the process of grinding food in the gizzard. Substitute lead for stone, and the bird has a problem. One pellet can kill. A further disastrous scenario is when the salts, oxides or hydroxides of lead can dissolve in water. Recognising this, the US government has banned the use of lead shot for hunting water birds since 1991, a position that is supported by the hunter and conservation group Ducks Unlimited and that is also the law in other countries such as Canada

Extending the lead ban to the uplands and to hunting rifles has been more difficult, but good progress is being made, especially in California, where a second wipeout of the tenuously conserved condors was attributed to the ingestion of bullets left in the abandoned carcasses of deer that lived long enough to escape and die later in a place remote to humans but not to condors. Hunters are now asked to bury the viscera stripped from killed game as it may contain slugs or bullet fragments. The same lead-ingestion risks apply to all carrion-eaters and raptors, including bald eagles. (As well as being affected by lead poisoning, bald eagles have been found to contain elevated levels of copper, since lead rifle bullets are frequently copper jacketed.) The recent suspicions concerning the continuing role of lead in condor mortality were confirmed when dangerously high blood levels were found, with some desperately sick birds being held for a time so that they could be rehabilitated by treatment with chelating agents.

Hunters are not the only culprits. Fishing sinkers are also a problem, with lead poisoning from this source being considered a major cause of death in the adult common loon, both in the northern USA and in Canada. Levels of greater than 5.0 ppm have been recorded in the livers of loons with sinkers caught up in their gizzards, compared with less than 0.1 ppm in those that lack these additional weights. Lead sinkers have been illegal in the United Kingdom since 1987, and Canada now bans the use of those weighing less than 50 grams.

Like raptors, we are predators, though our hunting and scavenging nowadays may be done in a supermarket or butcher's shop, or by perusing the menu at an up-market restaurant. Being human, I'm sometimes smart and sometimes very stupid. Until I read up on this lead shot story, it hadn't occurred to me that there might be a risk in eating venison, pigeon or partridge. Now I need to know whether my game order was from an animal killed in the wild and, if so, what type of shot was used. Somehow, it just seems easier and safer to stay with the range beef or the rack of lamb.

As the journalist and author John Moir has pointed out, there can be a lot of metal fragments in commercially sold meat that has been harvested by hunting. At least from my personal viewpoint, those who have been analysing elevated lead levels in the blood of carrion-eaters and investigating the underlying causes raise an important public health issue. We have been warned, and the warning has come from the sentinel condors and eagles that sample hunted game in the wild.

17

Red knots and crab eggs

VISITING THE CANADIAN PARLIAMENT in Ottawa as part of the 50th birthday celebration for the Toronto-based Gairdner International Awards for Medical Research, I was among several guests to be shown a magnificent original collection of Audubon bird prints in the elegant restored parliamentary library. Not long returned from the conservators, the collection mostly consisted of paintings that John James Audubon had made in Canada. Seeing those wonderfully coloured images in such pristine condition started me thinking about how the current situation for birds—which many believe is degrading rapidly with habitat destruction and anthropogenic climate change—can be compared with what was

happening in earlier times. Where do we get that historical data
for birds?

The Audubon paintings, and the comparable work that was
done in the Himalayas, Australia and Papua New Guinea by John
Gould and his associates, indeed provide an invaluable documented, historical resource. But it occurred to me that those great
nineteenth-century collections that long constituted many of the
central displays in natural history museums around the world must
also be of enormous importance. So when I found myself back
in Toronto and with some time off from the Gairdner anniversary celebrations, I arranged to meet with Allan Baker, the curator
of birds at the Royal Ontario Museum (ROM). Originally from
New Zealand, Allan is both an active researcher and a prominent
figure in the ornithology world.

Arriving at the ROM rear entrance, I was soon taken through
a portal to discover the mysteries that are back behind the familiar
public display rooms. Though I'm not a museum scientist, I felt
completely at home: state-of-the-art molecular biology research
laboratories were interfaced with corridors, offices and a meeting
room that, apart from the odd computer, appeared to be straight
out of the early twentieth century, or earlier. Less familiar, though,
were the rooms housing the bird collections. They held what I
most wanted to see and discuss.

The line of thought that started with my respectful viewing of
the Audubon prints had rapidly progressed to wondering whether
any preserved museum collections would be in good enough shape
to allow genetic comparisons between eighteenth- and nineteenth-
century specimens and contemporary birds. Furthermore, if the
material was still there, were people thinking seriously about
using it to do such 'molecular archaeology'? Allan Baker quickly
put me right, and I was delighted to hear that the answer was in
the affirmative for all cases.

First, though most of the old bird specimens were no longer
on display, they are being protected throughout the museum

world. Allan showed me drawers full of meticulously catalogued, arsenic-preserved bird 'skins', the standard method used by nineteenth-century collectors and museum conservators. Second, the material to hand was generally in sufficiently good condition to allow the recovery (by PCR) of this 'old' DNA with sufficient integrity to allow sequence analysis. Third, people were thinking about this, but the usual problem, as with most natural history investigations, was money. Humanity spends billions of dollars producing useless bombers that end up sitting in places like the Arizona desert, but we can't find the much smaller amounts needed to study nature in all its wonders and to ask how what's happening now may have implications for the future of life on this planet.

As we talked, Allan told me that most of the DNA work they'd been doing so far had focused on the genetic relationships between and within contemporary bird populations. Then, chatting more generally, he launched into an account of his own work. That's when I first heard about what has been happening with red knots and crab eggs. Allan also gave me a couple of reprints, including a general overview that had been published in the ROM magazine. Then, later, when I'd arrived back home, I downloaded some of his research papers and found that I was reading a detailed, data-driven analysis of the effects of human behaviour on a migratory bird population. Rigorously investigated and beautifully done, this is the story I will now lay out briefly for you.

A medium-sized sandpiper around 25 centimetres long and 50 centimetres or so in wingspan, the migratory red knot breeds right across the top of Canada, Russia and Europe, then heads down to Australia, New Zealand, South Africa and South America. Overall, the numbers are large, and the species globally is currently in the 'least concern' basket, but as you'll read below, that doesn't necessarily apply to regional populations. Their annual migration takes the knots from around 50°N to as far as 58°S, allowing them to spend much of their life in summer regions where food is abundant.

While the red knot's plumage is normally a modest, uniform pale grey, the bird takes the first part of its common name from the cinnamon-red colour of the head, throat, breast and belly that both males and females display in the breeding season. That rust-red theme is only slightly more prominent in the male, and this shared identity also comes through behaviourally in that both sexes incubate the eggs. Migrating flocks sound a 'knutt'-like grunting, which is where the 'knot' presumably comes from. The aspiring male makes a 'poor me' call in the mating season, a strategy that has long worked to ensure the female participation necessary for the continuation of many vertebrate species, including humans.

The *Calidris canutus rufa* subspecies studied by Allan Baker and his colleagues makes a marathon, 30 000-kilometre migration each year, flying from the Canadian Arctic deserts to Tierra del Fuego, then back again to breed in the northern summer. *C. c. rufa* red knots live for seven or eight years, travelling more than 400 000 kilometres in a lifetime. Each long trek involves only a couple of stops, so they put on considerable body mass before embarkation and must refuel along the way. Prior to setting out, migratory birds that fly very long distances can just about double their weight, from 90–120 grams to 180–220 grams for the *rufa* red knots. As they fly, fat provides the main energy store. Once that's gone, though, the birds start to burn protein from body organs like the liver and kidneys, the smooth muscle of the gizzard and intestines, then the striated pectoral muscles that enable flight, leading to a progressive decrease in overall strength. Well-supplied 'flyway' food outlets are clearly essential. On arrival, the integrity of the gastrointestinal tract and digestive system must be quickly reconstituted; otherwise they are more susceptible to attack by invading gut microorganisms.

The *rufa* knots are being studied intensively at several widely dispersed sites: the Canadian Arctic; the Mingan Archipelago on the north side of the Gulf of Saint Laurence; the southern extreme in Tierra del Fuego, and the Argentinian wetlands of San Antonio

Oeste, some 1400 kilometres to the north; and finally and perhaps most importantly, the Delaware Bay in the Philadelphia–Baltimore–Chesapeake region. The birds stop off to refuel at Mingan as they head south. The majority then travel the full distance to Tierra del Fuego, though some smaller groups prefer a more 'tropical' sojourn and end their southern trek in northern Brazil or Florida. Most of the red knots and great knots that come down to Australia from Eurasia also tend to finish their journey in the warmer north-west of the country, whereas the equally energetic short-tailed shearwaters (mutton-birds) that we see near our home travel from the Aleutian Islands and the Kamchatka Peninsula to end their southern pilgrimage on the islands and foreshores of chilly Bass Strait. Exhibiting the converse of the red knot profile, the mutton-birds lay and hatch their eggs in the south, then enjoy their non-breeding season in the north.

How do we work out what's happening with a migrating bird species? An obvious approach is just to count the numbers at different sites. A 2003 analysis of 207 shorebird populations showed that about half were declining while only a sixth were increasing. Then there's the more interventionist approach that Allan and his team used to study the red knots. The birds are first trapped using cannon nets, then handled gently while attaching a leg band or, more recently, coloured 'flags' with laser-encoded inscriptions that can be read through a high-powered spotting scope. A cannon net is just that: a large net is tied to projectiles that are then fired from smooth bore guns, with as many as four canons being used simultaneously to spread the net up and over before it settles gently to trap the birds. This is evidently well tolerated, and some handling is, of course, unavoidable if the birds are to be weighed so that their general body condition can be assessed and correlated with later survival.

Though there is ultimately plenty to eat during the far northern breeding season, the male red knots, in particular, need to arrive in reasonably good shape as they can reach the tundra before there is any food around. Also, the weather can still be very bad, and many die from the cold. That makes the Delaware Bay experience critical, as it is non-stop from there to the Arctic. During the five days or so before their May departure date, a bird can—providing the food supply is adequate—put on as much as 5–10 grams (average 4.6 grams) of fat and protein per day. That compares with only 1–2 grams per day at San Antonio Oeste for birds that come all the way up from Tierra del Fuego. Even the smaller population that travels only to Brazil still has to fly more than 5000 kilometres on the return trip to their mid-Atlantic stopover. Those reaching the Delaware Bay early gain the 47 grams plus of body weight that's needed to see them through the 2400-kilometre journey to the Canadian tundra and are spotted again in subsequent years, while late-arriving, low-weight birds tend to disappear permanently from view.

From 1997–98 to 2001–02, however, the proportion of well-conditioned red knots (200 grams or more) decreased by 70%. This was accompanied by a greater than 50% reduction in bird numbers. To understand what was happening, we need to take a closer look at the food chain. The very rapid bulking-up that enables the red knots to increase their fat and protein reserves with such enormous speed—the equivalent of 'super-sizing' at McDonald's—is achieved principally by gorging on the high-energy source of horseshoe crab eggs. Horseshoe crabs are commonplace on the Atlantic seaboard of the USA, from the Carolinas north, but these odd creatures will be unfamiliar to most who live in other parts of the world.

The fossil record suggests that the horseshoe crab (*Limulus polyphemus*) is one of the oldest creatures on the planet and that it hasn't changed much for more than 400 million years. Accustomed to the meat-laden blue swimmers (sand crabs) and Queensland mud (mangrove) crabs of my childhood, I did a double-take when I first saw these living fossils, with their small bodies, massive helmet-like carapace, spidery legs and armoured tails on the Long

Island beach near the famous Cold Spring Harbor Laboratory. Led for many years by Jim Watson, of Watson and Crick fame, Cold Spring Harbor has long been one of the main 'homes' of molecular biology, so even the most obsessed and narrowly focused 'bench' biologist will have visited there and will be able to recognise a horseshoe crab. Some have even worked with them.

Closer to the arachnids (spiders, ticks, scorpions) than the crab family, this blue-blooded invertebrate—they use the blue, copper-containing haemocyanin rather than the red, iron-containing haemoglobin to carry oxygen—has been used in vision research, while lysates of their white blood cells provide what has been the standard diagnostic test for detecting bacterial endotoxin. A cause of lethal shock in humans, endotoxin can also be a big problem for various cell culture systems. The limulus test depends on a unique molecule in horseshoe crab blood cells called coagulogen, which when released into fluid phase, aggregates in the presence of endotoxin. In nature, this mechanism no doubt works to help protect the crab against the consequences of bacterial infection. Technicians 'harvest' blood from crabs they find on the beach and then return them to the water. The endotoxin effect was discovered many years back by Johns Hopkins University scientist Fred Bang, an intellectually incisive and affable researcher who also did important early work in my field of viral immunity. I met him way back, at the beginning of my research career.

Horseshoe crabs have survived through the eons by producing a superabundant supply of eggs, so the sudden decline in red knot numbers was perplexing. The problem is, it seems, economics—otherwise known as human behaviour. Habitat change has likely contributed, but the main factor seems to be that, while horseshoe crabs have been used as bait for years, the numbers harvested between 1997 and 2002 for the conch, whelk and eel fisheries were massively increased. This led to a sixfold decline in horseshoe crab counts and a drastic drop in the availability of eggs to provide the equivalent of a tankful of 'Jet A' at the red knot fuelling depot. Estimates of population size for loggerhead turtles, the crab's principal predator, also fell dramatically. So far, pressures from the fishing community and the vagaries of the courts have meant that only New Jersey at the northern end of the Delaware Bay system has been able to sustain a total ban on horseshoe crab harvesting. One measure that helps reconcile the economic and conservation imperatives is to put the crabs used for whelk and conch bait in mesh bags, thus limiting their wastage due to attack by other species. Another is to use the crabs that have been bled for the limulus test.

But what led to this sudden increase in whelk harvesting? Whelk meat, which did not previously attract much of a price, suddenly became more valuable due to the declining availability of the Caribbean queen conch. That in turn resulted from massive overfishing during the preceding decade. Then concerns about the Delaware and Chesapeake blue swimmer populations led to regulations that drove some fishermen out of that market and into the conch, whelk and eel business. We value what we eat. Anyone who has dined on heavily spiced blue swimmers, washed down with beer in the unpretentious and unforgettable ambience of a Chesapeake or Delaware crab shack, understands why this culture has its advocates, including me. But nobody eats horseshoe crabs, so they don't have as wide a support base. And, of course, red knots aren't on any menu. The birds need about 50 000 horseshoe crab eggs per square metre in the top 5 centimetres of Delaware Bay sand if they are to do well, which translates to a catch of about 15 crabs per random net trawl. As of 2008, that number was down in the 1–5 range.

The case of the *rufa* red knots has been well publicised in the USA, and there is even a PBS documentary—though no more than 20% of Americans ever tune in to PBS. Still, establishing and implementing the regulatory changes that are needed to restore the horseshoe crab census to reasonable levels is a continuing battle. Repeat counts in Tierra del Fuego showed *rufa* red knot numbers declining from 53 000 in 2000, to 27 000 in 2002, to 14 800 in 2008, leading to a Canadian recommendation for endangered species status. Even so, the scientists are seeing some encouraging signs, with more returning juveniles being detected in Tierra del Fuego.

The basic message is that any sudden, human-induced change has consequences when it comes to natural (or harvested) eco-systems, which is just one of many reasons why widespread ignorance of science and the resultant failure to grapple with reality are dangerous. Despite heavy lobbying by local and inter-national conservation groups, the municipality of San Antonio Oeste recently excavated a tidal swimming pool in the most pro-ductive part of the *rufa* red knot wetland feeding area. It's too easy, though, to blame those in the poorer regions of the planet, when they are just replicating what we've long been doing.

Coastal wetlands are being threatened everywhere, by the construction of golf courses, housing developments, resorts and even automobile manufacturing plants. Oil spills also take their toll. Protecting the increasingly scarce wetlands that nurture both local species and visiting bird populations requires continuing action and vigilance. In an increasingly globalised world, how do we develop strategies to conserve key natural resources that tran-scend the boundaries of nation states? Birds, and fish for that matter, know no borders. That's also increasingly the case for the indigenous fishermen turned pirates in North Africa and Indonesia who, like some seabirds, have had their sustenance taken away by the over-exploitation, and even destruction, of fragile natural resources and habitats.

18

Hot birds

PHENOLOGY: THAT'S WHERE WE need to go next. You don't know what that is? Well, neither did I before I started to read up on this fascinating area of environmental science—a topic that is so different from my research field of immunology and infectious disease. The spell-checker on my word processing program underlines 'phenology' in red, though it accepts the dubious pseudo-science of 'phrenology' (reading skull bumps). The 1996 *Pocket Oxford Dictionary* that sits to the right of my computer goes straight from 'phenol' to 'phenomenal' with nothing in between. Then to Google, that arbiter of immediate resort, where I find that phenology is 'the study of periodic plant and animal life cycle events and how these are influenced by seasonal and inter-annual variations in climate'.

Other books—such as Møller, Fiedler and Berthold's *Effects of Climate Change on Birds*—examine far more comprehensively what's happening to avian species as the world warms, but we can't ignore climate change in this present exploration of the complex interactions between birds and humans. Realising that I was more than a bit out of my depth, I sought the advice of Lynda Chambers from the Australian Bureau of Meteorology. This is her research subject, and while there are many scientists focusing on phenology in the northern hemisphere, Lynda is one of the small group of professional researchers who grapple with the impact of climate change on those avian species that inhabit, or visit, the bottom half of our planet. As she is very quick to point out, her task would be impossible if it were not for the dedication of informed ama-teur observers from organisations like BirdLife Australia, the local equivalent of the US Audubon Society, who provide much of the data needed to track the distribution, numbers and movements of various avian species.

This isn't a book about anthropogenic climate change, but given the massive and well-funded disinformation campaign that's succeeded in confusing so many, it's necessary to point out a few basics. First, there's nothing we can do to influence many of the events that influence the climate cycle. It's obvious to everyone that we can't control either volcanic activity or the periodic changes in the obliquity of the earth's rotation that brings us closer to or further from the sun (thereby either heating or cooling our atmos-phere). Those 'Milankovitch cycles' largely determine the successive eons of glaciation and of ice-melting and rising oceans that charac-terise the story of this planet through geological time. That process continues, of course, but the precautionary principle that should dictate good governance requires that we act to mitigate green-house gas emissions. Though it may not be possible to attribute any single, isolated climate event to anthropogenic warming, the increased prevalence of extreme occurrences is exactly what climate scientists have been predicting for more than a decade. If you think that this is just academic hot air, watch what happens to insurance premiums for beachfront or 'tree-change' properties over the next few years—that is if it's possible to get any insurance coverage. Money talks.

Take, for instance, the situation in 2010: at least 1500 people died, and more than a million were displaced, when monsoonal rains caused unprecedented inundation in Pakistan. Extensive flooding in southern China and Central Europe caused enormous economic damage and some loss of life. In Russia, a historic heat wave was responsible for extensive crop failures; the forests burned; nuclear facilities were threatened, and Moscow choked on smoke and smog. Forest fires in 13 of Portugal's 18 provinces were the worst in living memory. Then, at the beginning of 2011, 800 000 people were displaced by the floods in Sri Lanka; 600 died in Brazil; and much of my home state of Queensland was shown to be one vast flood plain as torrential rains swept away crops, homes, people and infrastructure. Added to that, Yasi, a category 5 cyclone, caused the worst damage to north-eastern Australia that has been seen since human settlement, though coastal inhabitants were lucky in that the storm did not make landfall at full tide. In line with the idea that the warming oceans are putting more moisture into the atmosphere, both Europe and North America experienced massive snowfalls for the second successive winter. We still hear about '100-year events', but all the signs suggest that this is no longer a useful description.

When it comes to managing ever-greater ambient temperatures, the only place we can hope to change the equation is by manipulating the heat-trapping greenhouse gas effect. That was identified by Swedish chemist Svante Arrhenius in 1896, when he stated that 'if the quantity of carbonic acid increases in geometric progression, the augmentation of the temperature will increase nearly in arithmetic progression'. For carbonic acid, read CO_2, though other gases like nitrous oxide (N_2O) and methane (CH_4) are also of major importance. His initial calculation suggested that a doubling of atmospheric CO_2 levels would lead to a global mean temperature increase of 5–6°C, though he later revised that down to more like 2°C.

Atmospheric CO_2 levels measured at the remote Cape Grim site in Tasmania have gone from 330 ppm in 1975 to 385 ppm in 2010, representing a 1.6 ppm increase per annum. The numbers vary depending on winds, bushfires and so forth, but the cleanest days now are equivalent to the dirtiest recorded back then. For the northern hemisphere, annual increases in CO_2 at the Mauna Loa observatory in Hawaii went from 1.10 ppm in 1974–75 to 1.96 ppm in 2008–09 (with a range of 0.72–2.55).

All but a very small percentage of currently practising, well-published climate scientists are adamant that ever-increasing atmospheric CO_2 levels are dangerous. I've yet to read any recent research or discussion paper by a serious research investigator—including the few active climate scientists on the denier end of the sceptic spectrum—that suggests the link Arrhenius made between greenhouse gas accumulation and global temperature is fundamentally incorrect. Overturning Arrhenius would require us to rewrite the laws of physics.

The greenhouse gases that trap heat from the sun keep the surface of our planet from turning into an iceblock. Atmospheric CO_2 is, of course, essential for photosynthesis in plants, the basis of the food chain that fuels all vertebrate life. The problem is the enormous amounts of CO_2 that have been (and are being) released into the atmosphere as we extract and then burn the extraordinary reserves of fossil fuel that were laid down in the earth more than 300 million years ago. Ever-increasing methane levels from the rapidly melting permafrost in the far north are also contributing to the problem. So long as a box of matches is to hand, there's no longer any need to freeze on the tundra. As Dan

Miller shows on YouTube, poking a hole in the ice over a Siberian lake now accesses enough released methane to produce a steady, warming flame.

Based on computer modelling of scenarios ranging from 'low' to 'high', the prediction in the very conservative 2007 Intergovernmental Panel on Climate Change (IPCC) Synthesis Report for Policy Makers, which is signed off on by the representatives of the 300 or so member nations, is for a 1.8–4.0°C global mean temperature increase over the course of the twenty-first century. Most active researchers in the field of climate science are expecting, at a minimum, a 0.2°C rise per decade.

Anyone can look online at 'state of the climate' reports from a variety of responsible agencies, including the Australian Bureau of Meteorology and the US National Oceanographic and Atmospheric Administration (NOAA). According to NOAA, despite the somewhat cooler temperatures brought about by the La Niña conditions operating in Australia, the mean temperatures of the earth's land masses and oceans through 2010 were equal to those for 2005, the hottest year since 1880, when NOAA's National Climate Data Center began to take accurate measurements with standardised instrumentation. The inescapable, overwhelming reality is that all life forms, including birds, will have to deal with ever more frequent periods of extreme heat stress.

Obviously, many birds live in the tropics and others at (or near) the ice caps, while some, like the Arctic tern, roam the planet from pole to pole. Particular species have thus adapted to handle wide variations in ambient temperature. Migrants crossing the Sahara in summer tend to fly at night and rest up during the day. An emperor penguin, with its heavy insulation of fat and well-developed mechanisms for minimising heat loss, would find the waters of the Coral Sea discomfortingly warm, while a toucan placed outside in an arctic winter could expect to survive about as long as a snowflake in hell. Despite that, both birds maintain a normal body temperature in the 38–39°C range, though the life of a toucan is undoubtedly more languid in every sense than that of an emperor penguin.

In a warming world, the problem for toucans living in their normal home environment will be to stay cool. Everyone is familiar with the characteristic massive orange toucan beak. The biggest of all toucans, the toco toucan, has the largest beak-to-body ratio of any bird. This has nothing to do with toucan sexual display, or with feeding and/or defence, as you might think; the explanation is that the toucan beak functions primarily as a radiator. The toucan cools itself, or conserves heat when needed, by controlling the extent of blood flow through the beak's intense superficial vascular network. We do that too, but with our skin instead of a beak, and humans also have specialised sweat glands.

Birds don't sweat, but they do lose a fair amount of heat through their bare faces, legs, feet and (in some species) featherless patches on the head. Holding the wings out enhances air circulation over stretches of skin that are plentifully supplied with small blood vessels. Depending on how they're arrayed relative to each other and to the body, the complex arrangement of feathers helps to keep heat in or out. White, reflective tails can be turned towards the sun, though that strategy may only work well on calm days.

But what about dark-coloured birds? Anyone who lives in the tropics knows that black cars are hot! Shouldn't black cockatoos, for example, be at a distinct disadvantage in comparison with their white counterparts when it comes to heat uptake? But birds aren't just solid, inert bodies. Wind speeds as low as 3 kilometres per hour reverse the equation, so that black birds are better able to lose heat when ambient temperatures are high. The cooling effect of water transpiration also comes into it, and just as it does for Bedouin women, being clad in black facilitates better convection (and thus heat loss) between skin and feathers—or body surface, undergarments and robe, as the case may be.

Though the superficial layer (stratum corneum) of exposed skin is more scaly (keratinised) in birds than in us, birds can still lose about 50% of their body water by transpiration through this outer coat (integument). Species that live in very hot places— like the house sparrow in the Sahara, and the tropical dusky antbird—have evidently adapted to align the fatty acids of the stratum corneum in ways that facilitate the transport of water from the blood and tissue fluids to the skin surface. That lipid response strategy makes me wonder about the extent to which avian species that move between different environments modify normal biochemical mechanisms to deal with altered conditions. We've all seen birds splashing in ponds, baths or puddles of water in order to lose heat by promoting evaporation from wet feathers and skin. When a thoroughly saturated parent flies back to the nest, this 'belly soaking' is also used to cool the young.

Then the wonderful flow-through avian lung, which passes so much air so quickly (as we saw in Chapter 2), facilitates the evaporation of moisture from the respiratory mucosa. Like dogs, birds pant, and at a very high rate. Holding the beak open helps achieve maximum ventilation. Some species also use 'gular fluttering', a 'fanning' process that depends on the rapid relaxation and contraction of the throat hyoid musculature (which moves the tongue, pharynx and larynx) to increase airflow over the well-vascularised areas of the oropharynx and upper respiratory tracts. The overall effect is essentially the same as sweating, the common theme being that (latent) heat is required for vaporisation. The water molecules in sweat or respiratory mucous are closer together than those dispersed in the atmosphere. Driving them apart takes energy (heat), so the process of achieving that fluid-to-vapour transition cools the body.

Still, birds are generally more susceptible than mammals when it comes to very hot conditions. More than 200 birds, including 150 of the endangered Carnaby's white-tailed black cockatoo, are thought to have died from heat stress at the Western Australian south coast towns of Hopetoun and Munglinup in early January 2010, when ambient temperatures reached 47–53°C. Hopetoun is not all that far from Esperance, but the pathologists and chemists could find no evidence of lead (or any other) poisoning (as discussed in Chapter 16). The previous January, the science journalist Narelle Towie had reported in *PerthNow* that following good rains and an abundant breeding season, thousands of juvenile budgerigars and zebra finch were found dead in 45°C temperatures near the Overlander Roadhouse, 200 kilometres south of Carnarvon in Western Australia.

Such events are not new. The heat-related deaths of large numbers of zebra finch have been recorded as far back as the 1930s in various regions of western and southern Australia. The question is whether this will become more common. Ambient temperatures can reach a level where physiological mechanisms just cannot cope, especially under conditions of water shortage. And there is the additional problem that incubating chick embryos are particularly sensitive to hyperthermia. Continuing increases in ambient temperature with more frequent episodes of extreme heat stress must inevitably drive some avian species to extinction.

As temperatures go up, small birds dwelling in desert areas will require a 150–200% increase in water availability if they are to survive. The general trend so far is that the dry regions of the planet are becoming drier and the wet regions wetter, with more snow in northern latitudes. The long-term outlook for birds in desert areas may not be great. Fossil evidence from an earlier interglacial period (about 120 000 years ago) indicates that avian extinctions were most severe in areas like south-western Egypt, where, as a consequence of monsoonal change, there was a severe and permanent loss of rainfall.

When it comes to birds and acute heat stress, we can also be assured that there will be intensive, well-funded research programs that look at more than one species. This is because high ambient temperatures will compromise millions of dollars' worth of commercial egg and meat production from chickens, turkeys, quail and so forth. Even those birds that are caged and in sheds don't have air conditioning, as that would be both a major expense and, until we develop feasible solar-powered cooling systems, a substantial source of CO_2 emissions. Looking at the science literature, we can see that some work has already been done to analyse blood corticosteroid levels, heat shock proteins and other indicators of stress in domestic birds exposed constantly to 'hot' conditions. There are, for instance, measureable effects on the levels of reproductive hormones in heat-affected birds, which can lead to decreased egg laying. And immunity can also be compromised, increasing a bird's susceptibility to various infections.

What will the consequences of anthropogenic climate change be for birds in the wild? Thinking a little, it quickly becomes obvious that asking this for 'birds' is no more useful than posing the same question for 'placental mammals'. Unless we happen to mismanage the situation so badly that we create some total vertebrate catastrophe—by, for example, destroying all the algae and phytoplankton that recycle much of the CO_2 to give O_2—the effects of progressively increasing temperatures on various avian species will be no more uniform than for bears, badgers and buffaloes.

It might seem that birds are more adaptable than terrestrial mammals in that they should not require intact ice or land 'corridors'. However, some species will not, or cannot, transit rivers, roadways, mountain ranges and urban development islands. Others adapt to shrinking ecological niches, like cool mountaintops, where they are essentially trapped. Then there are the 'sedentary' birds—like many in the comparatively warm climates of Australia and Africa—that just aren't in the habit of moving around much. As their supportive habitats can be widely dispersed, particularly in the drier parts of these continents, there are very real concerns about the long-term prospects for these species.

Other species are more mobile. The very readable *State of the Birds* report, published by the US Department of the Interior in 2010, summarises a compelling and still accumulating body of evidence that many avian species in the northern hemisphere are tending to move their winter range. The report notes that:

> Although many factors are known to drive range changes, results from the Christmas Bird Count (CBC) show that the warmer winters in recent decades have played an important role in shifting winter bird ranges to the north. CBC data from the mid-1960s through 2006 show that 170 (56%) of the 305 most widespread, regularly occurring species have shifted their ranges to the north, whereas only 71 species (23%) have shifted to the south and 64 species (21%) have not shifted significantly north or south.

This review of the North American experience is complemented by a spectrum of European studies. Based on data cumulated over two decades, for instance, Vincent Devictor and his team concluded that although many avian species are moving away from the equator, this might not be happening fast enough. They recorded an average 91-kilometre change for the birds, compared with a 273-kilometre shift in temperatures. And Francisco Pulido and Peter Berthold have found that the blackcap, a passerine species that breeds in the Palaearctic in summer and has a southern migration, is wintering closer to its breeding grounds. They speculate that this might quickly be reflected in the genetic characteristics of the species, as the southern German blackcaps that have for years shown a profile of shorter migration now tend to dominate the species. This 'adaptive evolution' may to some extent be dictated by the selective disadvantage that results from immature juveniles attempting migration too early.

As Chambers and her colleagues have noted in the journal *Emu*, the same process of a pole-ward shift in preferred habitat looks to be happening for a number of species in the southern hemisphere. Thirty years back, the crested pigeon was never seen in Melbourne and its surrounds, but this bird is now commonplace in the city, a shift that reflects a steady move to higher latitudes. In Western Australia, birds are progressively relocating their range to the cooler (and wetter) south-west corner of the state. The other effect of continued warming is that the arrival and departure times for migratory species are changing. Looking at a compilation of published literature, bird observer reports and personal observations from birdwatchers, Sydney scientist Linda Beaumont and her colleagues summarised the southward migration data for more than two dozen species in south-eastern Australia and concluded that there was an average advance of arrival time of 3.5 days per decade, compared with a delay in departure of 5.1 days per decade, with, again, the patterns varying for different species.

One great concern is that increasing global temperatures will lead to a disassociation between the timing of bird migrations and the springtime rise in food availability that promotes reproductive success. Working from Groningen in the Netherlands, Marcel Visser and his colleagues described how the survival of pied flycatchers was greatly compromised between 1987 and 2003, when early spring conditions led to their migration being mistimed in relation to the supply of the caterpillars they feed to their fledglings. The birds still arrived on their normal schedule, but the flycatchers that nested in areas where caterpillar numbers peaked too soon suffered a 90% decline, compared with only 10% for those populations where the food–breeding relationship had remained stable—good news for the caterpillars and the various insect species they produce, bad for the birds. Like any situation where there is rapid environmental change, there are likely to be winners and losers.

And there are other types of evidence that temperature change has a direct effect on birds. For instance, my literature search turned up several reports from both the northern and southern hemispheres that at least some birds are getting smaller. Way back in 1847, German biologist Christian Bergmann proposed that, everything else being equal, lower environmental temperatures will cause homeotherms (birds and mammals) to grow larger, as unit body mass is inversely related to heat loss, while the converse will be true as the climate warms.

Looking at how Bergmann's rule might play out over the past century for eight passerine species from south-eastern Australia, Janet Gardner and her colleagues from the Australian National University found that there was a consistent trend across all eight species, while four out of the eight showed a significant decrease in both body size and wingspan (losing 1.8–3.6% of wing length). The birds in question were all sedentary, small- to medium-sized insectivores, an avian class that looks to be declining globally. The team got their data for 100 years ago from bird skins held in Australian museum collections (see Chapter 17). Also, as for tree rings, the size of feather bands is a direct reflection of nutritional status. Possible confounding effects related to differences in food supply over the 100 or more years of the study were thus assessed by determining the 'band-widths' on those archived 'skins', then comparing that data with estimates of overall bird size based on wing-length measurements.

Although I'm a complete outsider when it comes to avian phenology, my overall sense of this increasingly important, rigorous and highly professional field of research is that its findings so far provide substantial cause for concern. There can be no doubt that many bird species are moving their habitat range and/ or altering their migration times. As with anything to do with anthropogenic climate change, the effects are varied and complex. Some of the negative consequences will result from simple heat stress, which compromises the birds, their eggs and their breeding patterns. Other problems will result from habitat degradation and decreased or rapidly changing patterns of food supply. Altered profiles of insect distribution and availability must inevitably be a factor. Episodes of extreme dryness, for example, will lead to there being fewer mosquitoes for consumption by insectivores, while increasing ambient temperatures will move these disease-carrying vectors further up into the hills (as discussed in Chapter 11).

Rising atmospheric and seawater CO_2 levels will also have negative consequences for some avian species due to changes that do not depend directly on Arrhenius's greenhouse effect. Increasing concentrations of ambient CO_2 can cause plants to make more toxins and less protein. This provides a measure of pest control that is, nonetheless, not great for insectivorous birds. And the progressive acidification that results from the oceans taking up more CO_2 will ultimately compromise the shell quality, and thus the viability, of molluscs like mussels, whelks and so forth, which in turn affects seabirds further up the food chain. On our beach walks, we watch the big Pacific gulls carry mussels up into the air, then open their beaks so that their 'pre-packaged lunch' smashes open on the exposed rocks below.

There's also the effect of ocean acidification—where H_2O and CO_2 combine to H_2CO_3 (carbonic acid)—on the corals themselves and on the abundant environments that the reefs provide for other sea creatures. And marine biologists are generally of the view that, apart from compromising calcification through the production of carbonic acid, a 2°C rise in ocean temperatures will cause widespread reef loss from coral bleaching. Anything that affects the viability of corals, fish and littoral marine organisms could obviously have a negative impact on the food supply for seabirds. In so many respects, our failure to limit CO_2 emissions causes everincreasing damage to the web of life that nurtures the birds and, ultimately, us. Is there a more important issue facing humanity?

19

For the birds, and for us

THE WEB OF LIFE sustains us, gratifies us and increasingly demands our care and attention. From every conceivable aspect, humanity stands to lose a great deal, perhaps everything, if we do not accept this obligation of stewardship for the extraordinary diversity of species that inhabit the natural world. There are no guarantees for the continuation of any bird or any mammal, even those that, like the pigeons, swallows and starlings, seem utterly commonplace.

The need to sustain the constantly increasing numbers of humans leads inevitably to ever-diminishing resources for the support of other complex life forms. Cutting down forests, paving over productive land systems, draining aquifers and wetlands, and putting more and more cumulative toxins into the water, soil and atmosphere cannot be the recipe for a bright future. Perhaps our greatest challenge is to achieve the compassionate control of human population size without invoking the demons of war, starvation and disease. That entails bringing the linked ideas of global equity and diminished consumption into the broader consciousness.

The placentals that we encounter in the course of our day-to-day lives are generally domesticated and depend on our patronage (such as dogs and cats), or are subject to our exploitation (such as cows and sheep). We travel, often at great expense, to view lions, elephants, whales and wombats in their natural habitats. When it comes to viewing land mammals, such experiences are increasingly restricted to tourism in reserves protected by armed rangers. But even those of us who live in the most densely populated urban landscape can see and hear birds in the wild every time we walk the streets and look to the sky, or whenever we stare out through an open window.

Like us, birds are warm-blooded vertebrates. Unlike us, they live directly in nature, or what we choose to leave of nature. Watching the birds tells us what is happening in the broader environment. We should observe closely. In the long run, their fate is likely to be our fate. Since the seventeenth century and perhaps before, still life painters have sometimes included a dead bird, maybe a sparrow, as a 'memento mori' (reminder of mortality) at the foot of a canvas depicting, for example, the opulence of a vase of gloriously coloured flowers or an extravagantly set table. Our current lifestyles, though, are beyond extravagance.

Hunted to extinction, the North American passenger pigeon was once, by all accounts, the world's most abundant bird. At the time of European settlement, these migratory pigeons were so numerous throughout the Midwestern and Eastern states of Canada and the USA that estimates of total population size ranged from 3 to 5 billion. As John J Audubon recounted, enormous flocks could darken the sky for hours, or even days, as they passed overhead. But passenger pigeons made good eating, and being a social species that crowded together and sought safety in numbers, they were particularly vulnerable to that ultimate intelligent predator, *Homo sapiens sapiens*. Because they nested in massive colonies, it was easy to net large numbers of birds, or to stupefy them with alcohol-soaked grain so that they could be killed without even expending a single shot. Tied to a perch with eyes sewn shut and wings flapping, a captive bird could summon companions in the thousands. Then—that poor, sad 'stool pigeon' having done its job—the 'harvesting' would begin.

Like the great auk (extinct by 1844) in the northern hemisphere and the paradise parrot (extinct by 1927) in Australia, the passenger pigeon (extinct by 1914) is lost to us, except as a museum specimen, or as a visual or literary image. Realising what was happening, concerned and aware citizens tried to protect them. Some of the US states passed legislation, but the laws were weak and unenforced. Evidently Americans in general assumed that, because of their numbers, passenger pigeons would always be around.

In the end, deforestation and infection no doubt played a part in their final obliteration. Asking 'Who killed the last passenger pigeon?' could well be a classic case of posing the wrong question. Indiscriminate hunting may have reduced their numbers to the extent that they could no longer survive the toll taken by foxes, wolves and the various challenges associated with living in nature. Captive breeding programs did not work. Widespread recognition that the birds were endangered just came too late to save the passenger pigeon.

But for almost a hundred years, the plight of the passenger pigeon was no big secret. American novelist James Fenimore Cooper provided an account of their slaughter in his 1823 novel *The Pioneers*. They've figured since in songs, science fiction novels, movies, an episode of *Star Trek* and poetry. Born in 1915, the year after the last passenger pigeon died in the Cincinnati zoo, the passionate Australian environmentalist and poet Judith Wright published 'Lament for Passenger Pigeons' in her *Birds* collection.

Apart from contributing to the human food supply—a role that, at least in the West, has largely been supplanted for wild species by pre-packaged chickens and self-basting turkeys—much of the immediate appeal that birds have for us is aesthetic. An Audubon painting shows the passenger pigeon as an attractive bird, with the more iridescent male having a rose tint to the breast and pale blue on the wings. Though we will never again view those massive flocks that Audubon described, can there be a sight more gratifying than formations of cranes, pelicans or rose-breasted cockatoos in full flight? And what could be more emblematic of the predator–prey relationship than a hunting eagle as it launches from an ice floe to circle, then dives to scoop its prey from chilly waters, or more dramatic than seeing a cormorant plunge vertically into the sea?

When it comes to art and aesthetics, which are (of course) human constructs, the essence of what it is to be a bird may best be expressed in ballet. Anyone who has even the slightest interest in dance will be familiar with the grace of Tchaikovsky's *Swan Lake* (1876) and the drama of Stravinsky's *Firebird* (1910).

That tradition of creating performance around bird themes continues. When I was in Cambridge to give an evening seminar, a friend organised for me to have lunch with Professor Nicola Clayton FRS, who studies bird behaviour. I was intrigued: I'd never met an avian psychologist before. Slim, youthful-looking and personable, Nicky described her research on the food-caching activities of corvids. Equally fascinating, though, was our after-lunch conversation. As the conversation drifted, she told me a little about her 'other' professional life. Apart from being a top scientist, Nicky Clayton is a dancer, choreographer and scientific adviser to London's Rambert Dance Company. A recent collaboration with the company's artistic director, Mark Baldwin, led to the creation of a new performance piece that celebrates the

150th anniversary of Charles Darwin's *On the Origin of Species*. In her own words:

> *The Comedy of Change* is a collaborative project that combines my knowledge of evolution and animal behaviour with my research on the cognitive capacities of corvids (the crows and jays) and my passion for dance. I have always been fascinated by the showy displays of clever birds and their extravagant dances but this opportunity has inspired me to think in new ways.

Aesthetics and science come together in many ways, of course, with some of the most accessible examples being drawn from the spectacular photographs that we have of microscopic and other natural structures. When you think about it, the link should be obvious, for the best way to get scientific ideas across to a broader audience is to use the visual media. Movement adds a powerful dimension. Even the simplest life forms move, towards food for example, by sensing chemical gradients. That's how our white blood cells find their 'targets', and how eyeless 'soldier' termites defend their nest. It takes a brain, though, a coordinating central nervous system, to dance. And that's where the two halves of Nicky's professional life become one. As a scientist, she analyses avian thought processes and memory. As an artist, she translates that understanding into patterns of movement that impact on our consciousness.

Nicky focuses on the extent to which birds are capable of

Nicky Clayton and her rooks.

foresight, and makes the point that, as there is no way that we can directly interrogate either birds or small children, what we learn from experiments with birds also has implications for our understanding of human consciousness. I've since read some of her published works, but the study she summarised for me that afternoon related to the analysis of food caching by western scrub jays. While the long-distance travellers, like arctic terns, mutton-birds and red knots, ensure a continuity of food supply by migrating to regions where it is always spring, summer or autumn, those species that stay put in the cool or temperate parts of the planet have to store (or cache) enough calories to see them through the non-productive cold months. An individual corvid can evidently hide up to 30 000 items when food is plentiful and, what's more, is able to find them again through the harshness of winter. Don't think of yourself as a birdbrain when you can't recall where you put your car keys. At least as far as memory is concerned, some birds obviously have infinitely better brains than we do.

Such caching behaviour establishes that birds can display a measure of foresight, but is this an innate behaviour or does it reflect an intelligent decision-making process? Nicky's classic experiment involved putting scrub jays in adjacent cages. Each knew that other scrub jays were close by as they could always be heard, but one 'arm' of the trial placed a curtain between the cages so that there was no possibility of visual contact, while the 'controls' could both see and hear each other. Under conditions where the 'neighbours' could only be heard, an individual supplied with grain and a tray of sand or pebbles would elect to bury the food quietly in the soft sand. On the other hand, the control 'cacher', operating in line of sight, would choose the tray of pebbles. Any thieving jay that later tried to retrieve the food item would thus have to make a noise while disturbing the pebble overlay. Clearly, the cacher made a conscious decision concerning the choice of a 'silent' or 'alarmed' hiding place. Implicit in that behaviour, is there also a further recognition that any 'observer' will recall the hiding place when food later becomes scarce?

Nicky Clayton isn't unique in bridging the arts and sciences, but there are relatively few people who achieve professional standing in these two very different areas of creativity. Many draw enormous pleasure and aesthetic satisfaction from birds. Some translate that into art via photography, oils or watercolours, or by composing traditional or electronic music. And there are poets. Micheline Morgan, an artist and poet, offers the following, which encapsulates how being conscious of birds can affect our sense of self:

> *In my dream*
> *I find myself*
> *as I would lie in bed,*
> *belly to bed, face down,*
> *arms and legs*
> *stretched out and*
> *flapping through the air,*
> *effortless,*
> *gliding just above*
> *the tree line,*
> *or a little below,*
> *until I wake up,*
> *not grasping*
> *why I cannot*
> *fly again.*

Like the birds, we dance. But without the help of a machine, we can fly only in imagination! Art, music and poetry all help the human spirit soar free of those routines and obligations that can determine too much of our lives.

And great science is as much about creativity, insight and enhancing our vision as it is about measurement and reason. From Nicky Clayton's exploration of foresight to studies of pecking order in chickens, and from Malpighi's dissection of the chick embryo to Ronald Ross's elucidation of malaria and Pasteur's development of the first consciously attenuated vaccine, the systematic but imaginative analysis of how birds live in the world has informed us and continues to provide evidence-based enlightenment. And apart from their contribution to the sciences and aesthetics, birds do a great deal of other things that benefit our wellbeing and the health of the ecosystems that we share.

We have discussed in chapters 1 and 15 how eagles and vultures operate a sanitation service to remove the soft tissues of dead bodies. Pigeons clean up the food scraps from our cities, while insectivorous birds keep myriad invertebrate species under some measure of control. Birds pollinate plants and move seeds around, both attached to their feathers and in the contents of their gastrointestinal tracts. Seabirds bring nutrients from the oceans to the land. Anyone who has ever shared the home range of an Australian brush turkey will be acutely aware of the enormous amounts of soil that these birds can move, and aerate, in the process of building their massive nests. Birds like the oxpecker harvest ticks and other parasites off the backs of buffalo, hippopotamus and rhinoceros. Could any of us conceive what it would be like to live in a world without birds?

As a professional biomedical researcher, among the many new insights I've gained during the course of writing this book is the realisation that the love of birds is one of the few things that can transform those who have no formal training in science into enthusiastic practitioners of science. Every dedicated birdwatcher and bird-bander who exercises due diligence and makes careful, systematic observations is functioning as a scientist and serving the obligation of science to understand, protect and make sense of the natural world. The professionals who seek to document and interpret the consequences of habitat degradation, anthropogenic climate change and so forth for bird survival and migration patterns would have little data to work with if it were not for the activities of such dedicated amateurs. If you're bored, enjoy the outdoors and don't know what to do with your weekends or retirement, buy a pair of good binoculars, join the local chapter of BirdLife Australia, the Royal Society for the Protection of Birds, the British Trust for Ornithology, the American Birding Association, the Audubon Society or the like and get involved.

And those who document the numbers and movements of birds are indeed involved in serious, global science. The eBird project, run by Steve Kelling of Cornell University's Laboratory of Ornithology, had, by 2010, fed more than 48 million such observations into the US National Science Foundation's TeraGrid supercomputing network. The eBird data is then analysed in the context of information from NASA's Terra and Aqua satellites, which record, for example, the extent of forest cover and the greening associated with spring. Over the longer term, the TeraGrid database will be invaluable for developing a much better understanding of how bird populations are being affected by climate change and habitat degradation, allowing the identification of those species that are declining, or even facing extinction. Hopefully, this will help mobilise public opinion to protect the birds.

Also, even if we don't have the time or inclination to become observers, there are small things that each and every one of us can do to make life more secure for the birds. In those times of extreme heat stress that are quickly becoming more common across the planet, it takes very little effort to ensure that our garden or apartment balcony has a bird bath that is plentifully supplied with water. Bird feeders, though, are more contentious, as they can play a part in spreading infection. Feeders need regular cleaning and disinfection, and it's probably a good idea to seek the advice of the local birding organisation before investing in one of these things. They've been a prominent focus for spreading salmonellosis and the appalling psittacine beak and feather disease, caused by a circovirus that destroys the bird's immune system.

There's nothing to stop psittacine beak and feather disease from extending its ecological range if, for example, someone smuggles an infected Indonesian or Australian parrot into the USA or Europe. That may be what happened with the recent US outbreak of West Nile virus discussed in Chapter 5. West Nile virus encephalitis is now a permanent fixture of the North American avian and human disease scene. Ignoring quarantine regulations clearly enabled the high pathogenicity H5N1 bird flu virus to move south in Asia. Similarly the illegal transport of birds has always been a factor in spreading Newcastle disease. A major cause of economic loss in domestic poultry and a continuing problem for wild species, this parainfluenza type 3 virus is closely related to the pathogens that cause croup in small children. People handling chickens with Newcastle disease can develop conjunctivitis or mild respiratory infections. If you know, or suspect, that someone is moving birds nationally or internationally without proper approvals, inform the relevant authorities. This is a dangerous, criminal activity.

As we exercise our duty of care and maintain awareness of what is happening with avian species, we also embark on processes of conserving the natural environment that are ultimately essential for our own survival. In the course of just living, the free-flying and readily observed birds sample the atmosphere, the oceans, the plants, the forests and even insect life. If any one of those is compromised, the first place that such effects may become obvious is in the health and numbers of birds, both within and between species. No military commander would post sentinels, then ignore their warning cry and fail to monitor their continued wellbeing. As a defensive strategy in the face of massive and unpredictable environmental change, we will do well to think in terms of closely watched birds.

NOTES

I SEARCHING FOR PUFFINS

Page 3: ... Penguin Books branded their kid-oriented series with the amiable puffin.

It's interesting that, back in 1935, a British publisher chose to call itself Penguin. Real penguins, the birds not the books, live in the southern hemisphere, while puffins are restricted to the northern half of the planet. Perhaps Penguin's lack of parochialism reflected the fact that, as late as 1940 (the year I was born), the British Empire spanned the globe. Puffins and penguins share the characteristic of a black and white front, and it's also the case that, though less obviously comfortable in that position than a penguin, puffins do stand somewhat upright. However, the two species aren't closely related, and the lack of biological similarity is even greater for puffins and pelicans, the symbol that the Penguin publishers chose for their more heavy-duty, non-fiction books. What links these three species is that they are all seabirds, and that they make a memorable visual impact.

The short-tailed shearwater that commutes annually from the Arctic to breed in rookeries on the nearby Australian coast is classified as *Puffinus tenuirostris*, but the real puffins belong to the genus *Fratercula*, not *Puffinus*. Even so, though less committed travellers than the shearwater or the Arctic tern, which also visits our coastline, puffins are far from sedentary. The tufted variety we were expecting to see in Alaska locates further offshore during the winter months, while some European puffins are known to migrate between Wales and the North Atlantic.

Page 6: ... it was their relatives, the pigeon guillemots, that were most severely affected.

The other big media story that featured (in 2008) dead or dying pelicans in Australia described a situation that was in no sense a consequence of human action. When very heavy rains inundate the massive (and often completely dry) salt pans of ancient lakes in central Australia, there is an explosion of life that brings migrating birds, particularly long-distance flyers like the pelicans, to breed and gorge on the surprisingly abundant food sources. Bird numbers increase massively. But, as the lakes and watercourses dry out again, those growing pelicans that don't leave soon enough starve, with the result that they lack the strength to lift off and fly to better homes. The TV footage was awful, but nature in the raw can be far from pretty, and this must have happened over and over through the ages. The lesson for us is a general one: the fate of those stranded pelicans reflects what happens when a trapped population outgrows its food supply.

Page 6: ... those big electricity-generating windmills.

It's possible we could limit some of that problem if we painted the windmills purple rather than white. From insects to bats and birds, flying creatures are evidently purple-averse.

2 DISTANT RELATIVES

Page 16: ... the final fragmentation of the ancient mega-continent of Gondwanaland.

Australia was still part of Gondwana until about 45 million years ago, when the inexorable movement of the tectonic plates led to the breaking of the bridge between Tasmania and Antarctica. South America is thought to have separated from west Antarctica about 30 million years ago. The net result was that the waters surrounding the Antarctic were no longer routed north to the tropics and, staying closer to the poles, became much cooler. Ambient temperatures dropped and much of Australia dried out, an effect that was exacerbated as one plate overlapping the other drove the New Guinea highlands upwards, creating a rain shadow for the land to the south. Australia is still moving of course, and though the far north is gradually dipping down into the sea while the south slowly rises, this island continent should reach the equator within about 20 million years.

3 CHICK EMBRYOS AND OTHER DEVELOPING LIFE FORMS

Page 22: ... a particularly easy process to follow visually for the developing chick embryo.

By the time I got to Patten's later (1927) text on the *Embryology of the Pig*, I'd pretty much taken on board the basic nature of vertebrate development. Pigs are placental mammals like us, though the nature of the placenta varies a lot from species to species, and the diffuse (pig) and discoid (human) varieties are quite dissimilar. Setting aside the nature of the nutrient supply and the particular fetal incubator that maintains the appropriate temperature for *in ovo* or *in utero* development, the basic principles of early differentiation and organogenesis are not that different for chickens, pigs and us.

Page 22: ... a respiratory infection caused by a coronavirus.

Human coronaviruses of different types are one of the many causes of the common cold, while the SARS epidemic of 2002–03 was caused by a previously unknown coronavirus that originally came from bats, then transmitted via Himalayan civet cats to people.

Page 24: ... their 'blueprint or barcode' is written in the nucleic acid (RNA or DNA) sequences that they carry with them from, say, birds to mosquitoes to humans ...

Even the large DNA viruses (100–200 genes), like the poxviruses and herpesviruses, are too small to be seen using a conventional light microscope (as distinct from an electron microscope). And size isn't everything. The RNA 'barcodes' of the influenza A viruses and HIV consist of eight or nine genes respectively, and we all know how dangerous those two can be. By contrast, the smallest bacteria (*Mycoplasma*) have around 460 genes.

Page 24: ... featured large, industrial egg incubators.

Clad in varnished wood, they had a tray of water at the bottom for humidification, a big fan for air circulation and a heating element to keep them at around 37–38°C. Mammals are homeotherms (constant temperature) organisms, of course, and the body of the mother maintains the fetus at the right temperature, a strategy that imposes all the incubation responsibilities of reproduction on the female of the species. Birds are less sexist in this regard, as newly laid eggs can be kept warm by a sitting male or female. Both industrially and for the purposes of science, the post-fertilisation nurturing that's normally the role of one or both bird parents can be provided by a suitably heated box. Except in the very late stages of fetal development, that's yet to be achieved for any placental mammals.

Page 25: ... Q fever (*Coxiella burnetii*), a common problem for abattoir workers that has now been minimised by the development of a protective vaccine.

The story with Q fever, according to the Adelaide bacteriologist Barry Marmion, is that medical microbiologist Ted Derrick, working in my home town of Brisbane, identified the disease and sent samples to FM (Mac) Burnet, the great Melbourne biomedical scientist who shared the 1960 Nobel Prize for Medicine with the English transplant biologist PB Medawar. Burnet isolated the bug, and with help from prominent US scientists Herald Cox and Rolla Dyer at the National Institutes of Health, Bethesda, it was soon classified and named after him. Burnet, later Sir Mac, was a very forceful character, and there's some sense that Derrick's contribution may have been undervalued, although the organism was called *C. burnetii* at Derrick's suggestion.

Page 25: ... a clear, spreading 'plaque' of virus growth.

During the first half of the twentieth century—without even a minimal understanding of the different chemical and serum 'factors' that have since been identified as growth promoters for isolated cells—tissue culture was somewhat of an arcane art. Since the 1950s and 1960s, though, those of us who work with viruses that infect vertebrates have been able to do this type of experiment with 'lawns' of mammalian or avian cells grown on plastic Petri dishes under a semi-solid agar overlay. With the influenza viruses, for example, we often use a continuous line of cultured dog cells (MDCK, or Madin-Darby canine kidney cells) grown in Dulbecco's minimal essential medium supplemented with fetal calf serum, added growth factors and nutrients. (Renato Dulbecco won a Nobel Prize for his work on cancer, but his name is most commonly associated with the clear nutrient 'soup' he developed for growing cell cultures.) The discovery of the right nutrients and the availability of antibiotics to control bacterial contamination have transformed vertebrate cell culture from a highly specialised activity requiring very clean environments and an extraordinary level of sterile technique to a standard laboratory procedure that any competent person can follow.

Page 26: … those analysed for bacteriophages grown on plated bacteria.
This was the 'string and brown paper' era of science—still very much in operation at the time I started my research career. It could all go wrong when,
for instance, someone forgot to check the egg incubator before the weekend
and you came in on Monday to hear the cheep of hatchlings. I expect that
even Burnet—or maybe his technicians, who arrived early—had that experience. Most young virologists aren't familiar with the smell of melting paraffin wax. Bunsen burners were used routinely to flame the top of glass bottles
for sterilisation, an approach that doesn't work too well with the neck of a
plastic flask. Instead, we work in hoods with barrier airflow, which keeps
any contaminating bacteria or fungi at bay. The needles, rubber bulbs and
dental drills might just be found at the back of a drawer or cupboard in an
old facility, but most of the scientists who are working today wouldn't know
what they are for. One thing we may have lost, to our detriment, in this new
'plastics out of a box' era is the understanding that it's important to deal
respectfully with everybody, particularly the 'ladies in the kitchen' who wash
the glassware. Residual soap or detergent could really wreck an experiment.

4 SENTINEL CHICKENS

Page 30: … beyond that, the continuity of the crown of England.
For centuries, the last, short trip of many English men and women, including two of Henry VIII's six queens (Anne Boleyn and Catherine Howard),
was to Tower Green. Earlier on, the severed heads were impaled on a pike
and, together with the headless or hanged corpses, were often left on display as a warning to potential democrats, troublemakers and malefactors.
Omnivorous in their eating habits, ravens are highly intelligent as birds go
and have generally done well by keeping close to human settlements. The
largest of the passerine (perching) birds, they would, in times gone by, always
be found on voluntary clean-up duty around any active gallows or execu-
tion site. This somewhat grisly task has contributed to the raven's rather
macabre reputation, expressed most famously in Edgar Allan Poe's poem
'The Raven', in which 'this grim, ghastly, gaunt and ominous bird of yore'
repeatedly utters a single menacing word: 'Nevermore'.
    Exploiting that synergy between the natural (or unnatural) process of
death and aerial scavengers is, in fact, one of the more ecologically sound
ways of disposing of human and other remains, at least where population

densities are low and/or people are accustomed to a relatively high level of ambient bad odour. In modern times, though, fulfilling this form of sanitary duty can also have its downside and involve real risks for the birds (as we will see in Chapter 15).

Page 32: ... the terrible pathogen that kills humans by a combination of haemorrhagic disease and liver destruction.

Yellow fever had long been a major problem in the Americas, occurring as far north as Philadelphia and Boston in the heat of summer, but it came particularly to the fore in the 1890s with the US invasion of Cuba. After the deaths of a number of soldiers, a team led by army surgeon Major Walter Reed did what would now be completely unacceptable human transmission studies to prove beyond any doubt that YFV is carried by mosquitoes, confirming the earlier suspicion of the Cuban researcher and physician Carlos Findlay. Reed's medical colleague Jesse Lazear died because he deliberately exposed himself to infected mosquitoes.

Building on the firm scientific basis established by Findlay, Lazear, Reed and their colleagues, authorities dealt immediately with the yellow fever problem by instituting measures to limit mosquito breeding. Pools of standing water were drained or filled. Open cisterns were fitted with taps and the tops screened, or the water was covered with a layer of kerosene. This was also the approach used in 1904 by Colonel William Gorgas when the USA, now led by President Theodore Roosevelt, took over the construction of the Panama Canal. Among the difficulties that ended the earlier canal-building efforts of the French was the loss of workers from yellow fever and malaria, both of which can be defeated by comprehensive mosquito-control programs. What happened with yellow fever provided the first major breakthrough in our understanding of an arbovirus infection, though birds aren't maintaining hosts for either YFV or human malaria.

Page 33: ... an achievement recognised by his 1951 Nobel Prize.

Anyone who is vaccinated against YFV today will still be receiving a variant of Theiler's vaccine. Those who travel in, say, West Africa without taking it would have to be more than a little insane. Currently, the World Health Organization (WHO) reports that 90% of the 200 000 annual cases of yellow fever occur in Africa, with about one in six of those who develop obvious symptoms dying of the infection. The problem here is economic. One of the Millennium Development Goals (MGD4) announced in the year

2000, for example, was to deliver all the currently available vaccines to every child throughout the world. If we were to do that, it would mean the effective end of deaths from diseases like yellow fever.

Page 33: ... the identity of the key species can be incredibly hard to nail down.

In general, though we are the maintaining hosts for the various strains of dengue, human beings are only incidentally infected with most arboviruses. The fact that some of us develop severe symptoms like polyarthritis or meningoencephalitis (inflammation of the brain and its surrounding membranes, with destruction of nerve cells, or neurons) is incidental to the survival of the pathogen in nature. Among the many questions asked by infectious disease specialists is why one individual dies while another who later develops high levels of circulating antibody suffers, at worst, a mild headache. Does this reflect differences in our genetic make-up, or is it, perhaps, related to some physiological effect such as our level of exercise when we are circulating virus in blood? Fever is one of nature's ways of telling us to slow down, so don't run a marathon if your body temperature is up. Before the causative agents like RRV were identified, many of these infections were characterised as 'PUOs', pyrexias (fevers) of unknown origin.

5 FALLING CROWS

Page 40: ... in a viremic traveller who was incubating the disease.

Given the availability of a local, susceptible insect species, the 'human traveller–airplane dissemination' of arboviruses can potentially happen anywhere at anytime. Infectious disease physicians in Perth have, for example, detected Chikungunya virus in the blood of tourists returning from nearby vacation resorts in Asia. Despite the fact that the appropriate mosquito vectors are present, Chikungunya has yet to 'make the break' and become established in Australia. My guess is that it's just a matter of time.

Page 41: ... stockpiles had been destroyed by the early 1970s.

Even so, dangerous precedents were set, and it is hard to control the activities of fanatics and delusional psychopaths, whether they are acting individually or as part of menacing organisations. Though arboviruses won't do the job, we were reminded in 2001 that anthrax is a very effective tool for bioterrorists: *Bacillus anthracis* normally survives in the soil and is highly resistant to environmental damage. Terrorists are aiming to create a climate of fear, so they don't care in the least if, for example, a letter leaking anthrax spores kills a completely innocent postal worker. In nature, birds are at low risk of contracting anthrax, though there are reports of the disease occurring in ostriches, crows and ducks. Also, vultures have been implicated in spreading the spores via their gastrointestinal tracts after, for example, stripping the carcass of a cow or an antelope that has died from anthrax.

Page 42: ... so vaccination is the better option.

Veterinary vaccines can be made relatively quickly and cheaply, but the rigorous safety checks and more careful control of the manufacturing processes mandated by regulatory authorities like the US Food and Drug Administration (FDA) for any human vaccine make the whole process infinitely slower and more expensive. There's also the question of whether we really need such a product. While WNV did cause a major scare in the USA and has continued to spread from 1999 till now, the incidence of both human and avian disease does seem to have fallen of late. There is certainly evidence of mutational change over the years, and it could be that the virus is attenuating to become less virulent. The peak years of national incidence so far have been 2003 and 2006, but time will tell if that reflects any real downward trend or is just a consequence of variations in rainfall or some other climate effect that modifies bird habitats and mosquito numbers.

Page 44: ... is considered to have caused at least 1100 deaths.

As is the case with many such infections, the deliberately immunosuppressed (for cancer therapy or organ transplantation) and the elderly are particularly at risk. The cytotoxic drug and irradiation treatments that kill cancer cells also damage our immune systems, which, even for completely healthy individuals, become increasingly compromised as the years roll by. That ageing-related defect can become particularly obvious when it comes to dealing with novel pathogens. As with the brain, our long-term immune 'memory' often works better than our capacity to deal with new situations.

6 TICKS, SHEEP, GROUSE AND THE GLORIOUS TWELFTH

Page 47: ... the culture still survives, particularly in the Scottish Highlands.

A chance conversation with Ian Frazer, the biomedical researcher who drove the science that led to the human papilloma virus vaccine that protects against cervical cancer (Chapter 13), gave me a snapshot of the grouse-shooting culture. During the years that Hugh and I were doing our louping-ill studies in sheep, Ian, who is quite bit younger and grew up in Edinburgh and Aberdeen, spent part of his summer vacations picking up a bit of cash as a beater on a grouse moor. Though, in days gone by, the job of walking forwards in line across the moor while waving a flag to drive the birds from cover was done by the rural poor, it had, by the 1960s, been largely taken over by schoolboys—and more recently by girls as well. Accommodated in primitive 'bothies', which are basically the houses Scots crofters inhabited during the eighteenth century, the adolescent beaters have the lowest ranking on the grouse moor, well below the golden retrievers that have the job of fetching the shot game.

After several years at this, Ian was promoted in the beater hierarchy to a position of real responsibility, which required him to carry a ram's horn (inherited from a hill-walking aunt). Like the Archangel Gabriel, but without quite so much portent (though hopefully more immediate consequences), Ian was required to blow his horn as a warning when the beaters came within shotgun range. At that call, every 'gun' switched direction, shooting to aim backwards, away from the butts and the beaters. The loaders standing behind them would also have needed to adjust their position with some alacrity. As Ian recalls, it was not uncommon for many of the wealthy sportsmen to still be a little drunk, or seriously hung over from the night before. With hundreds of birds flushed from cover, the total 'bag' could be as few as two for a whole party. Such lack of dedication to the stated purpose of what was, after all, a very expensive social outing was evidently more typical of English shooting parties. When Germans came across for the hunt, the experience was much more efficient, sober and lethal (for the birds).

Page 49: ... cater to totally different market sectors.
Grouse for dinner, like the fabled larks' tongues for breakfast, may belong more to the pre–World War II era of Britain's 'great houses'. While there are published grouse recipes, from 'creamed grouse on toast' to 'oyster-stuffed grouse', I don't ever recall being in a restaurant that provided the opportunity of tasting grouse (although I believe Rules, the very old and very famous London game restaurant that was a favourite of the poet John Betjeman, does serve grouse). Grouse are evidently pretty tough birds and are best 'hung' in a cool environment for about two weeks prior to preparation for the table. Finding a suitable place isn't a problem in Scotland, where ambient temperatures are often too low for human comfort, but it's hard to ignore the fact that the tenderising effect of 'hanging' is a direct

consequence of tissue decomposition. The result is likely to be a very 'gamey' taste, which is not to the liking of everyone and is especially distasteful to most Americans.

7 FLU FLIES

Page 56: ... and that it was very small.

Virus diameters were measured by the capacity to pass as an infectious entity through the finest grades of sintered porcelain filter 'candles' made with different pore sizes by the Berkefeld Company, which is still in existence. In days gone by, the terms 'filterable' and 'virus' went together like 'gin and tonic' or 'Watson and Crick'. Some of those rough-surfaced, cylindrical white filters were still to be found in laboratories when I started studying viruses back in the 1960s. Now we could use synthetic membranes to make such size determinations, but there are many other ways to identify viruses and the main reason to do membrane filtration is to take out any larger contaminating microorganisms that would destroy the tissue cultures used for virus isolation. Samples recovered at post-mortem, for instance, may not be very clean.

Page 57: ... to recognise the contribution made by the humble ferret.

This little story also reveals why the virologists of the 1930s embraced the chick embryo inoculation techniques developed by Goodpasture, Woodruff and Burnet. Birds dispensed with teeth somewhere in the course of evolution. As Emile Zola reminded us, pigs bite and can kill. Have you ever tried to pick up an irritable ferret? Parrots and raptors are also best treated with caution, but nobody has ever been bitten to the bone by any form of chicken, especially an embryonic one. Fertilised hen's eggs are cleaner, cheaper and easier to house and handle than an angry mammal or wild bird. Science generally advances with the availability of more convenient and more sensitive methods and analytical systems, a process that continues to gather momentum.

Page 58: ... discovered in 1951 by Margaret Edney (later Sabine) while working as a very junior scientist in the laboratory of FM Burnet.

Margaret Sabine later switched from human to veterinary virology and was a much-loved teacher at the University of Sydney for many years. Among her research papers are publications entitled 'Paw and mouth disease in a cat', 'Feline picornavirus [polio-like] infections in Cheetahs', 'Towards a vaccine against equine herpesvirus 1', and 'Laboratory diagnosis of psittacine beak and feather disease by hemagglutination and hemagglutination inhibition (HI)'.

Page 58: ... and was soon well established in research and diagnostic laboratories.

Burnet evidently believed that the Swedes might have awarded him a Nobel Prize earlier (before 1960) for his influenza experiments if he'd beaten Hirst to the HI discovery. Science is both amiable and very competitive, and that was true even back then, when far fewer people were involved.

Page 59: ... it's a simple test to do and it is also very sensitive.

As with any technique in science, there are a few potential difficulties. One is that some serum samples also contain non-specific inhibitors that can mimic the effect of specific HI antibody, but these can be removed by prior treatment with another protein, called receptor-destroying enzyme (RDE), made from the bacterium *Vibrio cholerae*.

Page 59: ... can be read by measuring serum antibody titres.

The normal function of the flu surface H protein is to bind sialic acids (sugars) on the surface of susceptible respiratory epithelium, thereby initiating the process that leads to the virus gaining entry to the cell interior. The outer coat of the virus is then removed to expose the viral nucleic acid (RNA) information template. That initiates the virus growth cycle so that more virus can be made in the 'production factory' that might just happen to be one of our lung cells. At the end of this replication phase, the neuraminidase (N) protein on the surface of the virus operates to let the newly made infectious particles detach by cleaving the interaction between the H protein and the cell-surface sialic acid. The cell is severely damaged during the course of this reproduction process, and if the flu virus particles are not to die with it, they have to be freed so that they can ultimately be coughed or sneezed into the surrounding atmosphere.

Page 61: ... same method that's used to identify rapists and establish paternity.

The PCR story is told in Kary's entertaining biography, *Dancing Naked in the Mind Field*, an enlightening read if you think that all scientists are goody-two-shoes zombies in white coats. He shared the 1993 Nobel Prize for Chemistry with another friend, the late Mike Smith, who discovered how to change genes at will by site-directed mutagenesis. Kary's book cover has him posed with a surfboard, and Mike was a blue water sailor.

8 BIRD FLU: FROM HONG KONG TO QINGHAI LAKE AND BEYOND

Page 75: ... only to return again during the cooler months.

Much of what follows about the H5N1 viruses in the remainder of this chapter has been taken from a review article by the University of Hong Kong team of Hui-Ling Yen, Guan Li and Malik Peiris, with input from Rob Webster.

Page 75: ... and to 'Wallace's line' ...

Alfred Russell Wallace, the nineteenth-century co-discoverer (with Charles Darwin) of natural selection and evolution, supported himself for many years by trapping insects, birds and other wildlife in the East Indies, then selling them back to both amateur collectors and museums in the northern hemisphere. Many of the great museum collections of birds that we value today reflect the efforts of courageous and resilient human beings like Wallace. Much more than a trapper, embalmer and trader, Wallace had a keen intellect, which allowed him both to appreciate the mutability of species that is central to evolutionary theory and to analyse the distribution patterns of the different life forms he encountered.

9 BIRD FLU GUYS

Page 79: ... my two friends and colleagues Rob Webster and (the regrettably late) Graeme Laver ...

Some of what I recount here is taken directly from Graeme's obituary, which Rob wrote for the *Biographical Memoirs* of the Royal Society. I first met these two characters when I returned from the Moredun Research Institute, Edinburgh, to work at the John Curtin School of Medical Research (JCSMR) in the Australian National University, Canberra. The JCSMR Microbiology department, first headed by Frank Fenner, then by Gordon Ada, was an internationally renowned centre for research in virology.

Page 84: ... noddy tern virus that yielded the most spectacular crystals.

That analysis was further refined by solving the structure for co-crystals of neuraminidase bound to monoclonal antibody, one of the very first complexes between two different proteins to be defined in this way. Those experiments also involved Rob Webster, who, following techniques learnt from my Wistar Institute research collaborator, the influenza immunologist Walter Gerhard, had made the monoclonals.

Page 84: ... the Holy Grail for the pharmaceutical industry.
A later derivative, oseltamivir (Tamiflu), was a greater commercial success because it can be taken orally rather than by 'puffer', as is the case with Relenza. But zanamivir binds more tightly to the virus, and there is less mutational escape, making it the more effective drug. Laver, Colman and von Itzstein shared the 1996 Australia Science Prize for the development of Relenza.

10 BUG DETECTIVES

Page 87: ... visualised with the much greater power of electron microscopy.
The EM was co-invented in 1931 by Germans Ernst Ruska and Max Knoll. The first practical instrument was developed in 1938, but it wasn't until the 1950s that the technology advanced to the stage that we could see what influenza viruses look like as they bud from the surface of infected cells. Knoll died in 1969, and Ruska shared the 1986 Nobel Prize for Physics. If you want accolades in science, it's best to aim for a long life. Writing his Nobel Lecture (available on the Nobel e-Museum website) in the first decade of the twentieth century, Sir Ronald Ross described how he was able to work out the machinations of the malaria protozoan using a combination of clever experimentation and optical microscopy, but went on to say that nobody had been able to see the agent that causes yellow fever, though, like malaria, it had been shown to be transmitted by mosquitoes. Yellow fever virus is even smaller than flu.

Page 88: ... rather than from the infestation point of view.
Keats seemed particularly partial to crickets, mentioning them in his 'Ode to Autumn', then again in 'The Cricket and the Grasshopper'. Being a romantic who lived in gentler European landscapes, Keats was not, of course, think-ing of the plagues of grasshoppers that devastate crops and strip the minimal greenery from harsher environments like those of Africa and Australia. Shelley also wrote a little about crickets, but when it comes to larger flying organisms, he's better known for his bird poems:

Hail to thee blithe spirit,
Bird thou never wert—
That from heaven or near it,
Pourest thy full heart
In profuse strains of unpremeditated art.

None of us would ever regard 'To a Skylark' as an accurate biological description of *Alauda arvensis*, but the words are eminently satisfactory, reflecting, perhaps, the fact that human beings have viewed the natural world through the prism of beauty and the spirit for much longer than they have made systematic observations by peering down a binocular microscope. Even now, we much prefer Shelley's imagery to the idea that the skylark is carrying a gutful of bacteria.

Page 88: … were the last poets who knew anything much of science.
That's not strictly true, as there are more recent scientist-poets like the Czech Miroslav Holub and, as we'll read later in this chapter, the malaria hero Ronald Ross also tried his hand at verse. Perhaps Shelley, Keats, Coleridge and their contemporary William Blake were the last major poets who were able to encompass the scope of science, as it was in the first decades of the nineteenth century. Coleridge, for instance, knew Humphrey Davy, the great chemist and inventor of the miner's safety lamp.

Page 92: … enjoyed extensive naming rights.
Robert Koch's assistant, Julius Petri, is immortalised by the Petri dish, the round lab plates that, half-filled with agar, provide a solid medium favourable to the growth of many organisms. Nocard added haemoglobin to make the red blood-agar plates that are still used in many diagnostic microbiology laboratories. Then there's chocolate agar, so called because of the brown colour that reflects the presence of disrupted red blood cells heated to 56°C. Chocolate agar provides the nutritional requirements for growing some 'fussy' respiratory bugs, like the bacterium (not the virus) *Haemophilus influenzae*, which was briefly thought to be the cause of influenza. If you're getting the idea that this is all just cookery, then you aren't too far off. The original insight came from Fanny Hesse, who, working as a technician for her husband in Robert Koch's laboratory, was aware of the use of agar in jam-making. Back then, though, women didn't get naming rights in biology labs, and there are no Hesse plates.

An intriguing experience for those with an interest in the history of biology is to visit the museum of the Institut Pasteur at 25, rue du Docteur Roux in Paris. Pasteur, who became a hero in his time, is entombed in the basement of the building, constructed for him in 1888 by a grateful French people. For the last seven years of his life, he and his family lived in an apartment across the corridor from the laboratories where much of the equipment and some of the preparations made during his own lifetime are on display. You can, for instance, see an example of the delicate 'swan-necked' flask used to disprove the then current idea that infection is caused by some 'vital principle' in the air. The broth inside was boiled, and though the neck of the flask remained open to the surrounding atmosphere (but at lower level than the culture medium), the liquid remained clear and uncontaminated. Repeat the same experiment by leaving an open-topped jam jar on a tabletop for 24 hours or so and you will see rapid clouding due to the growth of airborne bacteria and other bugs that fall into the nutrient soup. Depending on what is around, it can soon look and smell very bad.

Page 92: ... the credit for the discovery goes to Smith.
Smith and Salmon made other important discoveries. Working together, they established that killed *S. enterica* could be used to make an effective vaccine. If we are wise, we have a similar product injected into our arms today before visiting any country where the water supply is not safe and typhoid is a possibility. Salmon also showed that bovine tuberculosis could be transmitted to humans, a major reason for the pasteurisation of milk and the campaigns that have eradicated TB from cattle in countries like Australia.

An 1891 investigation with the veterinarian Fred Kilbourne led Smith to identify the cause of bovine redwater, the tick-borne protozoan that we now call *Babesia*. One of my first professional jobs was as part of a weekend 'rotation' to screen blood smears from cattle injected with an 'attenuated' *Babesia* vaccine. When the numbers of red cells carrying the characteristic black dots shot up and the animals developed fever, they were immediately treated with a drug that killed off the parasite but left them solidly immune. A similar 'infection and treatment' approach was used in Africa to control another protozoan infection, bovine theileriosis, and could, perhaps, be applied to human malaria.

Page 95: ... and would later transmit that same parasite to clean sparrows.
A more detailed description of Ross's elegant studies in birds is to be found in his Nobel Lecture, which is easily accessed online. Apart from the science, it's

worth looking at as a sociological document. Ross lays out the difficulties he encountered and the progression of his thinking. He also conveys that sense of excitement and impatience that drives all good scientists. I found what he wrote way back then to be both delightful and very familiar.

11 HAWAIIAN WIPEOUT

Page 98: ... without putting humans at direct risk.

How much can just pass us by when we limit ourselves to a purely anthropo-centric world view! And I'm no exception. Despite working in the broad area of infection and immunity for almost 50 years, I hadn't even heard about the Hawaiian malaria problem until it came up in a chance conversation with a zoology colleague, Bob Day.

12 THE GREAT PARROT PANIC OF 1929–30

Page 106: The story was summarised in the *New Yorker* on 1 June 2009.

I am grateful to my St Jude colleague Chuck Sherr for directing me to Jill Lepore's fascinating article.

Page 106: ... the sequel to his more famous *Microbe Hunters*.

First published in 1926 and still in print, the *Microbe Hunters* is where most young scientists of my generation and before first encountered Louis Pasteur, Robert Koch, Emil von Behring and Theobald Smith. The book is not celebrated for its accuracy, and the chapter on Ronald Ross so infuriated its subject that it was withdrawn under threat of legal action. But de Kruif was an entertaining writer who also went on to tell the story of the early Dutch lens-maker and scientist Antony van Leeuwenhoek (1623–1724). Leeuwenhoek greatly improved the optical microscope and first described the submicroscopic world of 'animalcules', including spermatozoa and the bacteria that make up so much of our world's natural biota. Leeuwenhoek was the first real microbiologist, though the connection between free-living microorganisms like the ones he saw and infectious disease was not made until more than a century after his death.

Page 106: ... popular writers such as de Kruif and Sinclair Lewis ...

Lewis's 1925 novel *Arrowsmith* describes the life of young medical scientist and infectious disease specialist Martin Arrowsmith, who makes his way from a small town in the American Midwest to the laboratories of the prestigious

McGurk Institute, which everyone at the time (and now) regarded as synonymous with the Rockefeller Institute. *Arrowsmith* ends with the hero combating an outbreak of bubonic plague, caused by *Yersinia* (formerly *Pasteurella*) *pestis*. Awarded a 1926 Pulitzer Prize, the novel also describes how Martin Arrowsmith finds a solution to this disease in a therapy that, with the recent rise of antibiotic resistance, is again being discussed for bacterial infections. Arrowsmith's treatment involved the use of bacteriophages, viruses that can grow in and kill bacteria. It sounds good in principle, but it has yet to deliver the goods in clinical situations.

Page 106: People worried about 'germs' ...

Recently, we have been living with the possibility that our environment can be 'too clean' and with the realisation that children need some, hopefully fairly harmless, bugs in their lives if they are to avoid developing conditions like asthma. That doesn't include dangerous diseases like measles and whooping cough. It's essential to vaccinate kids against the severe infections of childhood, but let them play in the dirt a little and don't keep them too well scrubbed.

13 CATCHING CANCER

Page 111: ... cancer molecular biologists like my St Jude colleague Charles (Chuck) Sherr ...

Inspired at the beginning of his career by a visit to the laboratory of Mike Bishop and Harold Varmus (whom we'll meet later in this chapter), Chuck made the first of his several big cancer research 'hits' by analysing a virus-induced tumour, though his breakthrough had more to do with cats than with chickens.

Page 112: Cancer tends to be a disease of ageing ...

Though children get lots of infections and can develop various forms of cancer, paediatric tumours are both comparatively rare and tend to sort into a few well-defined categories where the underlying genetic causes are rapidly being unravelled. In fact, Mel Greaves at London's Imperial Cancer Research Fund's laboratories has been accumulating evidence that the normal infections (like snivels and croup) that all kids contract when they're sent to kindergarten or day care can help to set up their still-developing immune systems in ways that protect against some forms of leukemia.

Page 113: ... this discovery of what was soon called Shope fibroma virus ...

Over the years, we've realised that Shope fibroma virus (a poxvirus) is closely related to the pathogen that causes myxomatosis, the lethal infection that was introduced deliberately (in 1950) to control the Australian rabbit plague. Rabbits are much more valued in Europe, and vaccination with Shope fibroma virus was used as a protective measure to counter the myxomatosis epidemic that resulted from the deliberate introduction of this pathogen to kill off the rabbits on the estate of a prominent French physician. Though similar viruses can induce tumours in hares, squirrels and some monkey species, the horrible diseases associated with classical poxvirus infections of both chickens (avipox) and humans (smallpox) are not characterised by cancerous changes.

Page 116: The fibroblast technique produces ...

A familiar research tool for virologists and cell biologists of my era, chick embryo fibroblasts are made somewhat crudely by aseptically removing the embryo from the egg and forcing it through a sterile hypodermic syringe. The mishmash that emerges from the syringe nozzle is then dispersed in some suitable nutrient medium and incubated (at 37°C) in flat-sided tissue culture flasks or Petri dishes. Some of the cells then grow into monolayers. Exposure to a virus particle leads to the infection of a single cell then, after a few hours, the release of numerous progeny viruses, which, when dispersed in fluid phase, infect and then destroy the whole monolayer. Looking up at the plate using an inverted microscope (the objective is underneath and the light on top), all that may be seen is clear plastic and some floating cell debris. Adding a semi-soft agar overlay immediately after infection, though, prevents virus spread other than via direct cell-to-cell contact, allowing the emergence of clear, discrete 'plaques' that increase in size with successive cycles of virus replication and resultant cell death. Waiting a day or two, then counting the holes (plaques) in an otherwise pristine cell monolayer provides a direct measure of the number of input virus particles.

These tissue culture plaque assays took virology out of live animal systems (like Goodpasture and Burnet's chick embryos) and were at the cutting edge of the field in the mid-twentieth century. The approach can also be used to clone viruses and to measure titres of neutralising antibody. Various cell types were used, but if they supported the growth of the virus of interest, chick embryo fibroblasts were particularly popular as fertilised hen's eggs were cheap and readily available. Researchers never have enough money.

Page 117: ... for the Nobel Prize, which they shared with Renato Dulbecco in 1975.

Being awarded to a maximum of three people each year, Nobel Prizes in Medicine are as scarce as the proverbial hen's teeth. They mark major break-throughs in the biological sciences, yet here we have two (for Peyton Rous and Howard Temin) that depended on studies with the same little virus infection of domestic chickens.

Page 118: ... young French postdoctoral fellow by the name of Dominique Stehelin ...

The identification of *v-src* had depended on a finding by Martin, Duesberg (the later HIV/AIDS denier), Vogt, Hanafusa and colleagues, who defined an RSV mutant that was 15% shorter than the original 'wild-type' virus and that had lost the capacity to cause cancer, although it could still multiply. Reasoning the *v-src* would be contained in the genes of the missing 15%, Varmus and Bishop had Stehelin prepare a radioactive cDNA probe that was complementary to the sequences deleted from the non-cancerous RSV mutant. What they then found was that their *v-src* probe bound (annealed) tightly to DNA from normal cells, and from avian species as distant as the ratites.

Page 118: Bishop and Varmus, with the help of Vogt and Stehelin ...

Having done a key experiment, Dominique Stehelin was understandably upset when his name was missing from the 1989 announcement that the Nobel Prize for Physiology or Medicine had been awarded to Mike Bishop and Harold Varmus. The Nobel committees never comment, but their con-clusion was, presumably, that his contribution had been more technical and limited, and was made under the supervision of his senior colleagues. Still, life is not always fair, and that's often true for scientists and almost invariably the case for chickens.

14 BLUE BLOODS AND CHICKEN BUGS

Page 121: The idea that royalty and members of aristocratic European families were 'blue blooded' ...

Venous blood is, of course, 'exhausted' in the sense that it is oxygen-depleted, and aristocratic lineages can expire from a terminal lack of energy, exacer-bated by inbreeding, both social and physical. The latter is very apparent in some 'aristocratic' dog breeds, where a focus on the inheritance of a particu-lar combination of desired physical characteristics can lead to the selection of animals that are fundamentally 'unfit' in both the genetic and the functional sense, even to the extent that the animals suffer chronic pain.

Page 121: ... James I of England, also known as James VI of Scotland (1566–1625).

James was undoubtedly blue blooded, but he is also reported to have had blue urine, reflecting the presence of the pigment porphobilinogen, which would normally be broken down by the enzyme porphobilinogen deaminase. The disease then showed up again in George III's great-great-great grand-son Prince William of Gloucester (1941–1972), who was reliably diagnosed.

Page 121: ... Queen Victoria passed on the X-linked gene for haemophilia.

The haemophilia gene descended through Queen Victoria's daughter Alice (who wed Louis of Hesse), then through her daughter (also Alice), who married Nicholas II of Russia. The haemophilia of their son Alexei helped the mad mystic Grigori Rasputin to exert considerable influence over the royal family, creating tensions that ultimately played a part in triggering the 'glorious' revolution of 1917 and the subsequent mass murder of the Romanovs. The lesson is that mad mystics are best kept away from the seat of governance.

Page 122: ... to save soldiers wounded in the trenches of World War I.

After the war, suspecting that the parlous economic situation of Austria would not allow him to pursue his passion for medical research, Landsteiner moved to New York, and by 1923, he and Peyton Rous were colleagues at the Rockefeller Institute. Awarded a Nobel Prize in 1930, Landsteiner's other achievements were the first isolation of poliomyelitis virus (with Erwin Popper), the discovery of the human M, N and P blood groups, and the growth in culture of *Rickettsia prowazekii*, the bug that causes human typhus. There was so much to be discovered, and scientists were much less specialised back then

15 KILLING THE VULTURES

Page 130: ... Australian Animal Health Laboratory (AAHL) in Geelong ...

I saw this facility being built and was the first chair of its Scientific Advisory Committee. Recently, they've had major triumphs in identifying a number of completely novel fruit bat–borne infections, like the Hendra and Nipah viruses, which occasionally kill humans.

Page 131: Arnold Theiler ... is a genuine hero of infectious disease research.
In the early decades of the twentieth century, this Swiss-born veterinarian
founded the world-renowned Onderstepoort Laboratory, then the Pretoria
Veterinary School. Theiler was a great disease detective who identified a
number of infectious (and other) agents that kill African wildlife and livestock, and also developed some of the early veterinary vaccines. He has a
whole genus of nasty organisms named after him, particularly *Theileria
parva*, the protozoan parasite that causes East Coast fever in African cattle.
With a life cycle similar to that of malaria, this bug lives in the white rather
than the red blood cells, causing what is essentially a lethal form of leukosis.
Other *Theileria* species infect cattle in Australia and the Americas, but parasites like *T. mutans* and *T. buffelei* invade the red cells and are a much less
severe problem for animal production.

As usually happens on such occasions, Gerry Swan showed me around
what is a very impressive, well-equipped and modern veterinary school,
then he took me across the street to visit Arnold Theiler's Onderstepoort
Laboratory. The main building remains, and there is a statue of the great
man out the front. It was quite an experience—like stepping through a time
portal—to stand in his office and see the notebooks and the early, simple laboratory equipment that he and his associates used. Theiler's rooms have been
maintained in their original state, with lots of paper, polished woodwork,
teak, copper and glass, materials that have been replaced in contemporary
laboratories by computers, composites, plastics and stainless steel.

16 HEAVY METAL

Page 136: ... with the result that the birds are poisoned.
My only scientific paper on heavy metal toxicity resulted from just such an
occurrence with sheep, where the consequence was lethal liver and brain
damage (spongiform encephalopathy) comparable to that seen in human
Wilson's disease.

17 RED KNOTS AND CRAB EGGS

**Page 142: ... the 50th birthday celebration of the Toronto-based Gairdner
International Awards for Medical Research ...**
The Gairdners, as they're known, are one of the two national research
prizes that have a high success rate in predicting future Nobel Laureates in
Physiology or Medicine. That 'indicator' status, which also adds prestige to

the American Lasker Awards, shows that (like the Nobel juries) the selection committee are both informed and hard working. Our Ottawa visiting group consisted of four Gairdner–Nobel 'double threats': yours truly, my colleague Rolf Zinkernagel, Bengt Samuelsson and Harald zur Hausen. Bengt discovered the prostaglandins that are important for inflammation, and Harald worked out that human papilloma virus causes cervical cancer in women, which was the necessary foundation for Ian Frazer's development of a protective vaccine.

Page 142: … in the elegant restored parliamentary library.
To fill in some time between events, we were taken on a tour of the parliamentary library. Though the elaborate 1876 panelling in the central reading room had been seriously damaged in a fire many years ago that destroyed the old parliament building, prompt action by a librarian, who closed the connecting door, allowed the important books and records to be saved. Prominent among those were the Audubon bird prints.

Page 142: … the collection mostly consisted of paintings that John James Audubon had made in Canada.
In January 2011, an original of JJ Audubon's *Birds of America* sold for more than US$7 million.

Page 144: Allan showed me drawers full of meticulously catalogued, arsenic-preserved bird 'skins' …
Naively, having stared at stuffed animals since I was first taken on museum visits by my maternal grandfather at about the age of five, I was expecting to see taxidermy: whole mounted birds in various 'natural' postures. There were a few of those around the walls, and many others in the public displays that I visited later, but that was only a small part of the scientific collection. Most birds are preserved as round skins, with the numbers in some collections being in the hundreds of thousands. The body tissues and skeleton are first removed, then what remains is processed in some way to prevent bacterial decomposition and fading. In days gone by, the preferred treatment was with arsenic trioxide or sodium arsenite, meaning that such specimens must, of course, be handled with caution. The arsenic preserves the vivid colours and, fortunately, also leaves the DNA intact. Thinking about it further, I realised that old Indian headdresses, feather boas and the like may also be an important source of DNA, so those with such antiques might need to guard them carefully when enthusiastic research ornithologists are in the neighbourhood.

Page 146: ... enjoy their non-breeding season in the north.

Around the time of the shearwater arrival in the latter half of September, it's common to see a few dead birds washed up along the beach. In other months, we may see the sad remains of an occasional fairy penguin, though they're evidently thriving locally with warmer ocean temperatures. In November 2009, though, we encountered ten dead mutton-birds on our regular beach walk, then more than 50 the next day. Alarmed, we enquired and found out that such shearwater 'wrecks' have been happening for at least the past century. A possible explanation is that when the birds arrive from the north at the limits of their physical endurance, the weather is very bad or the fish supply is down for some reason and they are just unable to rebuild strength with sufficient speed. But is that really what has happened? Making a definitive analysis of a periodic and unpredictable event like this requires considerable resources, both financial and human, to sustain continuing, long-term monitoring, sampling and testing.

18 HOT BIRDS

Page 152: ... the massive and well-funded disinformation campaign that's succeeded in confusing so many ...

If you doubt the validity of the preceding assessment, then get hold of *Merchants of Doubt: How a handful of scientists obscured the truth from tobacco smoke to global warming*, by Naomi Oreskes and Erik Conway. Legitimate science is not in the business of deception and confusing the public. What makes this book so important is that the case is very carefully documented.

Page 155: ... mean temperatures of the earth's land masses and oceans through 2010 were equal to those for 2005 ...

The following is directly reproduced from NOAA's January 2011 report:
- For 2010, the combined global land and ocean surface temperature tied with 2005 as the warmest such period on record, at 0.62°C (1.12°F) above the 20th century average of 13.9°C (57.0°F). 1998 is the third warmest year-to-date on record, at 0.60°C (1.08°F) above the 20th century average.
- The 2010 Northern Hemisphere combined global land and ocean surface temperature was the warmest year on record, at 0.73°C (1.31°F) above the 20th century average. The 2010 Southern Hemisphere combined

global land and ocean surface temperature was the sixth warmest year on record, at 0.51°C (0.92°F) above the 20th century average.

- The global land surface temperature for 2010 tied with 2005 as the second warmest on record, at 0.96°C (1.73°F) above the 20th century average. The warmest such period on record occurred in 2007, at 0.99°C (1.78°F) above the 20th century average.
- The global ocean surface temperature for 2010 tied with 2005 as the third warmest on record, at 0.49°C (0.88°F) above the 20th century average.
- In 2010 there was a dramatic shift in the El Niño–Southern Oscillation, which influences temperature and precipitation patterns around the world. A moderate-to-strong El Niño at the beginning of the year transitioned to La Niña conditions by July. At the end of November, La Niña was moderate-to-strong.

19 FOR THE BIRDS, AND FOR US

Page 167: … more dramatic than seeing a cormorant plunge vertically into the sea?

This ability was, of course, harnessed in Asian countries by fastening a snare round the base of the cormorant's neck so that the bird could only swallow small fish. The fisherman would then 'help' the cormorant by removing those larger fish that got stuck in the throat. Like falconry, such traditional practices did little, if anything, to disturb the balance of nature.

LATIN BINOMIALS FOR COMMON BIRD NAMES

amakihi, *Hemignathus virens*
antbird, dusky, *Cercomacra tyrannina*
auk, great (extinct), *Pinguinus impennis*
auklet, rhinoceros, *Cerorhinca monocèrata*
bellbird, or bell miner, *Manorina melanophrys*
blackcap, *Sylvia atricapilla*
budgerigar, shell, or common pet parakeet, *Melopsittacus undulatus*
cassowary, species of *Casuarius*
chicken, domestic, *Gallus domesticus*
chukar, or chukar partridge, *Alectoris chukar*
cockatoo, black, *Calyptorhynchus funereus*
cockatoo, Carnaby's, or white-tailed black, *Calyptorhynchus latirostris*
cockatoo, rose-breasted, or galah, *Eolophus roseicapilla*
cockatoo, white, or sulphur-crested, *Cacatua galerita*
condor, California, *Gymnogyps californianus*
cormorant, great, *Phalacrocorax carbo*
cormorant, little pied, *Phalacrocorax melanoleucos*
crake, marsh birds of the family *Rallidae*, commonly genus *Porzana*
crane, black-necked, *Grus nigricollis*
crow, American, *Corvus brachyrhynchos*
crow, hooded, *Corvus cornix*
currawong, pied, *Strepera graculina*
eagle, bald, *Haliaeetus leucocephalus*
fantail, grey, *Rhipidura fuliginosa*
finch, Laysan, *Psittirostra cantans*

finch, zebra, *Taeniopygia guttata*
flamingo, Chilean, *Phoenicopterus chilensis*
flycatcher, pied, *Ficedula hypoleuca*
goose, bar-headed, *Anser indicus*
goose, Canada, *Canada granadensis*
grouse, red, *Lagopus lagopus scotica*
guillemot, pigeon, *Cepphus columba*
guineafowl, domestic, *Numida meleagris*
gull, black-headed, or Pallas's, *Larus ichthyaetus*
gull, Pacific, *Larus pacificus*
hen harrier, *Circus cyaneus*
heron, Nankeen night, or rufous night, *Nycticorax caledonicus*
heron, white-faced, *Egretta novaehollandiae*
honeyeater, New Holland, *Philiydonyris novaehollandiae*
jay, western scrub, *Aphelocoma californica*
knot, great, *Calidris tenuirostris*
knot, red, *Calidris canutus*
knot, red, subspecies *rufa*, *Calidris canutus rufa*
kookaburra, laughing, *Dacelo novaeguineae*
lark, passerines of the family *Alaudidae*
loon, common, *Gavia immer*
lorikeet, purple-crowned, *Glossopsitta porphyrocephala*
magpie, Australia, *Gymnorhina tibicen*
magpie, yellow-billed, *Pica nuttalli*
murrelet, marbled, *Brachyramphus marmoratus*
myna, common hill, *Gracula religiosa*
oxpecker, red-billed, *Buphagus erythrorhynchus*
oxpecker, yellow-billed, *Buphagus africanus*
oystercatcher, Icelandic, *Haematopus ostralegus*
parrot, orange-bellied, *Neophema chrysogaster*
parrot, paradise, *Psephotus pulcherrimus*
pelican, Australia, *Pelicanus conspicillatus*
pelican, brown, *Pelicanus occidentalis*
penguin, adélie, *Pygoscelis adeliae*
penguin, emperor, *Aptenodytes forsteri*
penguin, little, or fairy, *Eudyptula minor*
pheasant, domestic, *Phasianus colchicus*
pigeon, crested, *Ocyphaps lophotes*
pigeon, homing, or rock, *Columbia livia*

pigeon, passenger (extinct), *Ectopistes migratorius*
pochard, or diving duck, *Aythya* species
puffin, Atlantic, *Fratercula arctica*
puffin, horned, *Fratercula corniculata*
puffin, tufted, *Fratercula cirrhata*
quail, domestic, *Coturnix* species
quail, Japanese, *Coturnix japonica*
raven, common, or northern, *Corvus corax*
raven, little, *Corvus mellori*
rook, *Corvus frugilegus*
shearwater, short-tailed, or mutton-bird, *Puffinus tenuirostris*
shearwater, wedge-tailed, *Puffinus pacificus*
shelduck, ruddy, *Tadorna ferruginea*
skylark, *Alauda arvensis*
sparrow, house, *Passer domesticus*
starling, European, *Sturnus vulgaris*
swan, whooper, *Cygnus cygnus*
tern, Arctic, *Sterna paradisaea*
tern, black noddy, *Anous minutus*
tern, little (western Pacific), *Sterna albifrons sinensis*
toucan, toco, *Ramphastos toco*
toucans, tropical, *Ramphastos* of some eight species
turkey, Australian brush, *Alectura lathami*
turkey, domestic, *Meleagris gallopavo gallopavo*
turkey vulture, or buzzard, *Cathartes aura*
vulture, African bearded, *Gypaetus barbatus*
vulture, African white-backed, *Gyps africanus*
vulture, griffon, *Gyps fulvus*
vulture, Indian, *Gyps indicus*
vulture, Indian white-rumped, *Gyps bengalensis*
vulture, slender-billed (also in India), *Gyps tenuirostris*
vulture, white-backed, *Gyps africanus*
wattlebird, red, *Anthochaera carunculata*
weaverbird, passerines of the family *Ploceidae*
whipbird, eastern, *Psophodes olivaceus*

ABBREVIATIONS

AAHL	Australian Animal Health Laboratory, Geelong, Victoria
ABO	human blood group polysaccharides (sugars)
AFIP	Armed Forces Institute of Pathology (US)
B cell	lymphocyte, or white blood cell, that develops in the avian Bursa of Fabricius
B locus	avian MHC locus
BAL	British anti-Lewisite
CDC	Centers for Disease Control and Prevention (US), formerly the Communicable Diseases Center, Atlanta, Georgia
cDNA	complementary DNA
CJD	Creutzfeldt Jacob disease
DNA	deoxyribose nucleic acid
FDA	Food and Drug Administration, the US agency that licenses vaccines and drugs
FRS	Fellow of the Royal Society of London, the British National Academy of Sciences, and the oldest in the English-speaking world
H	haemagglutinin
H2-KDL	mouse class I MHC genetic loci
HLA-ABC	human lymphocyte class I MHC genetic loci
IUCN	International Union for the Conservation of Nature
JCSMR	John Curtin School of Medical Research
JEV	Japanese encephalitis virus
JFK	John F Kennedy International Airport, New York
LI	louping-ill virus
MDV	Marek's disease virus
MHC	major histocompatibility complex

MVE	Murray Valley encephalitis virus
NASA	National Aeronautics and Space Administration (USA)
NOAA	National Oceanographic and Atmospheric Administration (USA)
NSAID	non-steroidal anti-inflammatory drug
ppm	parts per million
RDE	receptor destroying enzyme
RNA	ribose nucleic acid
ROM	Royal Ontario Museum
RRV	Ross River virus
RSV	Rous sarcoma virus
TB	tuberculosis
T cell	lymphocyte, or white blood cell, that develops in the thymus
UN	United Nations
USPHS	United States Public Health Service
WEHI	Walter and Eliza Hall Institute
WHO	World Health Organization
WNV	West Nile virus
YFV	yellow fever virus

FURTHER READING AND REFERENCES

BACKGROUND READING

Andrewes, C & Pereira, HG 1972, *Viruses of Vertebrates*, 3rd edn, Balliere Tindall, London.

Birkhead, T 2008, *The Wisdom of Birds: An illustrated history of ornithology*, Bloomsbury, New York.

Chansigaud, V 2010, *All About Birds: A short illustrated history of ornithology*, Princeton University Press, Princeton & Oxford.

Greger, M 2006, *Bird Flu: A virus of our own hatching*, Lantern Books, New York.

Knipe, DM & Howley, PM (eds) 2001, *Fields Virology*, vols 1 & 2, 4th edn, Lippincott Willimas & Wilkins, Philadelphia.

Nobel e-Museum. A number of Nobel Laureates and their work are mentioned through the course of this book. Both their personal biographies and the lectures they gave at the time of the award can be accessed online at the Nobel e-Museum website: http://nobelprize.org/nobel_prizes/medicine/laureates/

Pizzey, G & Knight, K 2003, *The Field Guide to the Birds of Australia*, 7th edn, HarperCollins, Sydney.

Tate, P 2007, *Flights of Fancy: Birds in myth, legend and superstition*, Random House, London.

Watson, JD 2000, *A Passion for DNA: Genes, genomes and society*, Cold Spring Harbor Laboratory Press, New York.

I SEARCHING FOR PUFFINS: AN INTRODUCTION

Armstrong, RH et al. 1980, *A Guide to the Birds of Alaska*, Alaska Northwest Publishing Co., Anchorage.

Butchart, SHM et al. 2010, 'Global biodiversity: indicators of recent declines', *Science*, vol. 328, no. 5982, pp. 1164–8.

Goldsworthy, SD, Gales, RP, Giese, M & Brothers, N 2000, 'Effects of the *Iron Baron* oil spill on little penguins (*Eudyptula minor*) 1: estimates of mortality', *Wildlife Research*, vol. 27, no. 6, pp. 559–71.

Golet, GH et al. 2002, 'Long-term direct and indirect effects of the "Exxon Valdez" oil spill on pigeon guillemots in Prince William Sound, Alaska', *Marine Ecology Progress Series*, vol. 241, pp. 287–304.

2 DISTANT RELATIVES

Feduccia, A 1999, *The Origin and Evolution of Birds*, 2nd edn, Yale University Press, New Haven.

Maina, JN et al. 2010, 'Recent advances into understanding some aspects of the structure and function of mammalian and avian lungs', *Physiological and Biochemical Zoology*, vol. 83, no. 5, pp. 792–807.

Tudge, C 2009, *The Bird: A natural history of who birds are, where they came from and how they live*, Three Rivers Press, New York.

3 CHICK EMBRYOS AND OTHER DEVELOPING LIFE FORMS

Beveridge, WIB 1947, 'Simplified techniques for inoculating chick embryos and a means of avoiding egg white in vaccines', *Science*, vol. 106, no. 2753, pp. 324–5.

Burnet, FM 1969, *Changing Patterns: An atypical autobiography*, American Elsevier, New York.

Dunbrack, RL & Ramsay, MA 1989, 'The evolution of viviparity in amniote vertebrates: egg retention versus egg size reduction', *American Naturalist*, vol. 133, no. 1, pp. 138–48.

Goodpasture, EW, Woodruff, AM & Buddingh, GJ 1932, 'Vaccinal infection of the chorio-allantoic mebrane of the chick embryo', *American Journal of Pathology*, vol. 8, no. 3, pp. 271–82.

Marmion, BP and Feery, B 1980, 'Q-Fever', *Medical Journal of Australia*, vol. 2, pp. 281–2.

Patten, BM 1946, *The Early Embryology of the Chick*, 3rd edn, Blakiston, Philadelphia.

Sexton, C 1999, *Burnet: A life*, Oxford University Press, Melbourne.

4 SENTINEL CHICKENS

Doherty, RL 1974, Arthropod-borne viruses in Australia and their relation to infection and disease, *Progress in Medical Virology*, vol. 17, pp. 136–92.

Eron, C 1981, *The Virus that Ate Cannibals: Six great medical detective stories*, Macmillan, New York.

Fenner, FJ (ed.) 1990, *History of Microbiology in Australia*, Australian Society of Microbiology, Canberra.

Komar, N, Dohm, DJ, Turell, MJ & Spielman, A 1999, 'Eastern equine encephalitis virus in birds: relative competence of European starlings (*sturnus vulgaris*)', *American Journal of Tropical Medicine and Hygiene*, vol. 60, no. 3, pp. 387–91.

MacKenzie, JS & Williams, DT 2009, 'The zoonotic flaviviruses of Southern, South Eastern and Eastern Asia and Australasia: the potential for emergent viruses', *Zoonoses and Public Health*, vol. 56, nos 6–7, pp. 338–56.

Nathanson, N (ed.) 1997, *Viral Pathogenesis*, Lippincott-Raven, Philadelphia.

Peters, CJ & Olshaker, M 1998, *Virus Hunter: 30 years of battling hot viruses around the world*, Anchor Publishing, Canada.

Theiler, M & Downs, WG 1973, *The Arthropod-borne Viruses of Vertebrates: An account of the Rockefeller Foundation virus program 1951–1970*, Yale University Press, New Haven.

Van den Hurk, AF et al. 2010, 'Vector competence of Australian mosquitoes for Chikungunya virus', *Vector-Borne and Zoonotic Diseases*, vol. 10, no. 5, pp. 489–95.

5 FALLING CROWS

CDC, Division of Vector Borne Diseases, 'West Nile virus', www.cdc.gov/ncidod/dvbid/westnile/surv&control.htm#surveillance

Crosbie, SP et al. 2008, 'Early impact of West Nile virus on the yellow-billed magpie (*Pica nuttalli*)', *Auk*, vol. 125, no. 3, pp. 542–50.

Eidson, M et al. 2001, 'Crow deaths as a sentinel surveillance system for West Nile virus in the northeastern United States, 1999', *Emerging Infectious Diseases*, vol. 7, no. 4, pp. 615–20.

Ernest, HB, Woods, LW & Hoar, BR 2010, 'Pathology associated with West Nile virus infections in the yellow-billed magpie (*Pica nuttalli*): a California endemic bird', *Journal of Wildlife Diseases*, vol. 46, no. 2, pp. 401–8.

Kuno, G & Chang, G-J 2005, 'Biological transmission of arboviruses: reexamination of and new insights into components, mechanisms, and unique traits as well as their evolutionary trends', *Clinical Microbiology Review*, vol. 18, no. 4, pp. 608–37.

Southam, CM & Moore, AE 1951, 'West Nile, Ilheus and Bunyamwera virus infections in man', *American Journal of Tropical Medicine and Hygiene*, vol. 1, no. 6, pp. 724–41.

Work, TH, Hurlbut, HS & Tylor, RM 1955, 'Indigenous wild birds of the Nile Delta as potential West Nile virus circulating reservoirs', *American Journal of Tropical Medicine and Hygiene*, vol. 4, no. 5, pp. 872–8.

6 TICKS, SHEEP, GROUSE AND THE GLORIOUS TWELFTH

Gilbert, L et al. 2004, 'Ticks need not bite their red grouse hosts to infect them with louping ill virus', *Proceedings of the Royal Society B: Biological Sciences*, vol. 271, supplement 4, pp. S202–5.

Reid, HW 1999, 'Monks, ticks and the molecular clock', *Biologist*, vol. 46, pp. 197–200.

7 FLU FLIES

Andrewes, CH, Laidlaw, PP & Smith, W 1935, 'Influenza—observations on the recovery of virus from man and on the antibody content of human sera' *British Journal of Experimental Pathology*, vol. 16, no. 6, pp. 566–82.

Barry, JM 2004, *The Great Influenza: The story of the greatest pandemic in history*, Viking Press, New York.

Crosby, AW 2003, *America's Forgotten Pandemic: The influenza of 1918*, 2nd edn, Cambridge University Press, Cambridge.

Mullis, K 1998, *Dancing Naked in the Mind Field*, Pantheon, New York.

Neiderman, CS, Sarosi, GA & Glassroth, J (eds) 2001, *Respiratory Infections*, 2nd edn, Lippincott, Williams & Wilkins, Philadelphia.

Shope, RE 1931, 'The etiology of swine influenza', *Science*, vol. 73, pp. 214–15.

Taubenberger, JK, Hultin, JV & Morens, DM 2007, 'Discovery and characterization of the 1918 pandemic influenza virus in historical context',
Antiviral Therapy, vol. 12, no. 4 (part B), pp. 581–91.

Yuen, KY et al. 1998, 'Clinical features and rapid viral diagnosis of human
disease associated with avian influenza A H5N1 virus', *Lancet*, vol. 351,
no. 9191, pp. 467–71.

8 BIRD FLU: FROM HONG KONG TO QINGHAI LAKE AND BEYOND

Feare, CJ, Kato, T & Thomas, R 2010, 'Captive rearing and release of bar-
headed geese (*Anser indicus*) in China: a possible HPAI H5N1 virus
infection route to wild birds', *Journal of Wildlife Diseases*, vol. 46,
no. 4, pp. 1340–2.

Lebarbenchon, C, Feare, CJ, Renaud, F, Thomas, F & Gauthier-Clerc, M
2010, 'Persistence of highly pathogenic avian influenza viruses in natural
ecosystems', *Emerging Infectious Diseases*, vol. 16, pp. 1057–62.

Salomon, R & Webster, RG 2009, 'The influenza virus enigma', *Cell*,
vol. 136, no. 3, pp. 402–10.

Webby, RJ, Webster, RG & Richt, JA 2007, 'Influenza viruses in animal
wildlife populations', *Current Topics in Microbiology and Immunology*,
vol. 315, pp. 67–83.

Yen, H-L, Guan, Y, Peiris, M & Webster, RG 2008, 'H5N1 in Asia', in H-D
Klenk, MN Matrosovich & J Stech (eds), *Avian Influenza*, Monographs
in Virology, vol. 27, Karger, Basel, pp. 11–26.

9 BIRD FLU GUYS

Downie, JC, Webster, RG, Schild, GC, Dowdle, WR & Laver, WG 1973,
'Characterization and ecology of a type A influenza virus isolated from
a shearwater', *Bulletin of the World Health Organization*, vol. 49,
no. 6, pp. 559–66.

Mandavilli, A 2003, 'Profile: Robert Webster', *Nature Medicine*, vol. 9,
no 12, p. 1445.

Pereira, HG, Tumova, B & Webster, RG 1967, 'Antigenic relationships
between influenza A viruses of human and avian origins', *Nature*,
vol. 215, pp. 982–3.

Webster, RG 2010, 'William Graeme Laver', *Biographical Memoirs of
Fellows of the Royal Society*, vol. 56, pp. 215–36.

Von Itzstein, M 2007, 'The war against influenza: discovery and development of sialidase inhibitors', *Nature Reviews Drug Discovery*, vol. 6, pp. 967–73.

10 BUG DETECTIVES

de Kruif, P 1960, *Microbe Hunters*, Harcourt Brace, New York.

Feldman, B 2000, *The Nobel Prize: A history of genius controversy and prestige*, Arcade Publishing, New York.

Geison, GL 1995, *The Private Science of Louis Pasteur*, Princeton University Press, Princeton.

Holmes, R 2008, *The Age of Wonder*, Vintage Books, New York.

Nobel e-Museum website: http://nobelprize.org/nobel_prizes/medicine/laureates/.

11 HAWAIIAN WIPEOUT

Atkinson, CT, Dusek, RJ, Woods, KL & Iko, WM 2000, 'Pathogenicity of avian malaria in experimentally-infected Hawaii Amakihi', *Journal of Wildlife Diseases*, vol. 36, no. 2, pp. 197–204.

Beadell, JS et al. 2006, 'Global phylogeographic limits of Hawaii's avian malaria', *Proceedings of the Royal Society B: Biological Sciences*, vol. 273, no. 1604, pp. 2935–44.

Warner, RE 1968, 'The role of introduced diseases in the extinction of the endemic Hawaiian avifauna', *Condor*, vol. 70, pp. 101–20.

12 THE GREAT PARROT PANIC OF 1929–30

Armstrong, C 1933, 'Psittacosis', *American Journal of Nursing*, vol. 33, pp. 97–101.

Lepore, J 2009, 'It's spreading: outbreaks, media scares and the parrot panic of 1930', American Chronicles, *New Yorker*, 1 June, pp. 46–50.

Storz, J 1971, *Chlamydia and Chlamydia-induced Diseases*, Charles C Thomas, Springfield.

13 CATCHING CANCER

Biggs, PM 2004, 'Marek's disease: long and difficult beginnings', in F Davison & V Nair (eds), *Marek's Disease: An evolving problem*, Elsevier Academic Press, London, pp. 8–16.

Bishop, MJ 1989, 'Retroviruses and oncogenes II', Nobel Lecture, 8 December, http://nobelprize.org/nobel_prizes/medicine/laureates/1989/bishop-lecture.html, viewed March 2012.

Ridley, M 2006, *Francis Crick: Discoverer of the genetic code*, HarperCollins, New York.

Rous, P 1966, 'The challenge to man of the neoplastic cell', Nobel Lecture, 13 December, http://nobelprize.org/nobel_prizes/medicine/laureates/1966/rous-lecture.html, viewed March 2012.

Shope, RE 1932, 'A filterable virus causing a tumor-like condition in rabbits and its relationship to virus myxomatosum', *Journal of Experimental Medicine*, vol. 56, pp. 803–22.

Shope, RE 1935, 'Serial transmission of virus of infectious papillomatosis in domestic rabbits', *Proceedings of the Society for Experimental Biology and Medicine*, vol. 32, no. 6, pp. 830–2.

Temin, HW 1975, 'The DNA provirus hypothesis', Nobel Lecture, 12 December, http://nobelprize.org/nobel_prizes/medicine/laureates/1975/temin-lecture.html, viewed March 2012.

Tooze, J 1973, *Molecular Biology of Tumor Viruses*, Cold Spring Harbor Monograph, New York.

Varmus, HE 1989, 'Retroviruses and oncogenes I', Nobel Lecture, 8 December, http://nobelprize.org/nobel_prizes/medicine/laureates/1989/varmus-lecture.html, viewed March 2012.

14 BLUE BLOODS AND CHICKEN BUGS

Baxter, AG 2000, *Germ Warfare: Breakthroughs in immunology*, Allen & Unwin, Sydney.

Bumstead, N & Kaufman, J 2004, 'Genetic resistance to Marek's disease', in F Davison & V Nair (eds), *Marek's Disease: An evolving problem*, Elsevier Academic Press, London, pp. 112–23.

Briles, WE 1984, 'Early chicken blood group investigations', *Immunogenetics*, vol. 20, no. 3, pp. 217–26.

Garwin, L & Lincoln, T (eds) 2004, *A Century of Nature: Twenty-one discoveries that changed science and the world*, University of Chicago Press, Chicago.

Kaufman, J 2008, 'The avian MHC', in F Davison, B Kaspers & KA Schat (eds), *Avian Immunology*, Elsevier, London, pp. 161–83.

Lyall, J et al. 2011, 'Suppression of avian influenza transmission in genetically modified chickens', *Science*, vol. 331, no. 6014, pp. 132–3.

Mak, TW & Saunders, ME 2006, *The Immune Response: Basic and clinical principles*, Elsevier, New York.

15 KILLING THE VULTURES

Cardoso, M et al. 2005, 'Phylogenetic analysis of the DNA polymerase gene of a novel herpesvirus isolated from the Indian *Gyps* vulture', *Virus Genes*, vol. 30, no. 3, pp. 371–81.

Gutsche, T 1979, *There was a Man: The life and times of Sir Arnold Theiler KCMG of Onderstepoort*, Howard Timmins, Cape Town.

McGrath, S 2007, 'The Vanishing', *Smithsonian Magazine*, February, www.smithsonianmag.com/science-nature/vulture.html, viewed March 2012.

Naidoo, V & Swan, GE 2008, 'Diclofenac toxicity in *Gyps* vultures is associated with decreased uric acid excretion and not renal portal vasoconstriction', *Comparative Biochemistry and Physiology Part C: Toxicology and Pharmacology*, vol. 149, no. 3, pp. 269–74.

Oaks, JL et al. 2004, 'Diclofenac residues as the cause of vulture population decline in Pakistan', *Nature*, vol. 427, no. 6975, pp. 630–3.

Pain, DJ et al. 2008, 'The race to prevent the extinction of South Asian vultures', *Bird Conservation International*, vol. 18, pp. S30–48.

Swan, G et al. 2006, 'Removing the threat of diclofenac to critically endangered Asian vultures', *PLoS Biology*, vol. 4, no. 3, pp. 395–402.

16 HEAVY METAL

Carpenter, JW et al. 2003, 'Experimental lead poisoning in turkey vultures (*Cathartes aura*)', *Journal of Wildlife Diseases*, vol. 39, no. 1, pp. 96–104.

Massachusetts Department of Environmental Protection, Office of Research and Standards 11 May 2009, *Assessment of Ecological Risk Associated with Lead Shot at Trap, Skeet & Sporting Clays Ranges*, Massachusetts Department of Environmental Protection, Boston, www.mass.gov/dep/toxics/stypes/lsersk.pdf, viewed March 2012.

Moir, J 2006, *Return of the Condor: The race to save our largest bird from extinction*, Lyons Press, Guilford, CT.

Western Australia, Parliament, Legislative Assembly, Education and Health Standing Committee 2007, *Inquiry into the Cause and Extent of Lead Pollution in the Esperance Area*, November, Parliament of Western Australia, Perth.

17 RED KNOTS AND CRAB EGGS

Baker, AJ 2009, 'The plight of the red knot', *ROM Magazine,* vol. 41, pp. 18–23.

Baker, AJ et al. 2004, 'Rapid population decline in red knots: fitness consequences of decreased refuelling rates and late arrival in Delaware Bay', *Proceedings of the Royal Society B: Biological Sciences,* vol. 271, no. 1541, pp. 875–82.

Gonzalez, PM, Baker, AJ & Echave, ME 2006, 'Annual survival of red knots (*Calidris canutus rufa*) using the San Antoio Oeste stopover site is reduced by domino effects involving late arrival and food depletion in Delaware Bay', *Hornero,* vol. 21, no. 2, pp. 109–17.

18 HOT BIRDS

Australian Bureau of Meteorology website: www.bom.gov.au/climate/change/

Beaumont, LJ, McAllan, IAW & Hughes, L 2006, 'A matter of timing: changes in the first date of arrival and last date of departure of Australian migratory birds', *Global Change Biology,* vol. 12, pp. 1339–54.

Both, C, Bouwhuis, S, Lessells, CM & Visser, ME 2006, 'Climate change and population declines in a long-distance migratory bird', *Nature,* vol. 415, no. 7089, pp. 81–3.

Chambers, LE, Hughes, L & Weston, MA 2005, 'Climate change and its impact on Australia's avifauna', *Emu,* vol. 105, no. 1, pp. 1–20.

Chambers, LE 2008, 'Climate change and birds: a southern hemisphere perspective', *CAWCR Research Letters,* vol. 1, pp. 8–33.

Chambers, LE 2010, 'Altered timing of avian movements in a peri-urban environment and its relationship to climate', *Emu,* vol. 110, no. 1, pp. 48–53.

Cullen, JM, Chambers, LE, Coutin, PC & Dann, P 2009, 'Predicting onset and success of breeding in little penguins *Eudyptula minor* from ocean temperatures', Marine Ecology Progress Series, vol. 378, pp. 269–78.

Devictor, V, Julliard, R, Couvet, D & Jiguet, F 2008, 'Birds are tracking climate warming, but not fast enough', *Proceedings of the Royal Society B: Biological Sciences,* vol. 275, no. 1652, pp. 2743–8.

Gardner, JL, Heinshohn, R & Joseph, L 2009, 'Shifting latitudinal clines in avian body size correlate with global warming in Australian passerines', *Proceedings of the Royal Society B: Biological Sciences,* vol. 276, pp. 3845–52.

McKechnie, AE & Wolf, BO 2010, 'Climate change increases the likelihood of catastrophic avian mortality events during extreme heat waves', *Biology Letters*, vol. 6, no. 2, pp. 253–6.

Miller, D, *Fire and Ice: Permafrost Melt Spews Combustible Methane*, www.youtube.com/watch?v=1liqk9UQNAQ, viewed March 2012.

Munoz-Garcia, A, Ro, J, Brown, JC & Williams, JB 2008, 'Cutaneous water loss and sphingolipids in the stratum corneum of house sparrows, *Passer domesticus* L., from desert and mesic environments as determined by reversed phase high-performance liquid chromatography coupled with atmospheric pressure photospray ionization mass spectrometry', *Journal of Experimental Biology*, vol. 211, part 3, pp. 447–58.

National Oceanic and Atmospheric Administration (NOAA) website: http://www.ncdc.noaa.gov/sotc/global/

North American Bird Conservation Initiative, US Committee 2010, *The State of the Birds 2010: Report on Climate Change, United States of America*, US Department of the Interior, Washington, DC, www.stateofthebirds.org

Pulido, F & Berthold, P 2010, 'Current selection for lower migratory activity will drive the evolution of residency in a migratory bird population', *Proceedings of the National Academy of Sciences of the United States of America*, vol. 107, no. 16, pp. 7341–6.

Rozenboim, I, Tako, E, Gal-Garber, O, Proudman, JA & Uni, Z 2007, 'The effect of heat stress on ovarian function of laying hens', *Poultry Science*, vol. 86, pp. 1760–5.

Tattersall, GJ, Andrade, DV & Abe, AS 2009, 'Heat exchange from the Toucan bill reveals a controllable vascular thermal radiator', *Science*, vol. 325, no. 5939, pp. 468–70.

19 FOR THE BIRDS, AND FOR US

Grodzinski, U & Clayton, NS 2010, 'Problems faced by food-caching corvids and the evolution of cognitive solutions', *Philosophical Transactions of the Royal Society B: Biological Sciences*, vol. 365, no. 1542, pp. 977–87.

Marris, E 2010, 'Supercomputing for the birds', *Nature*, vol. 466, no. 7308, pp. 807.

Stulp, G, Emery, NJ, Verhulst, S & Clayton, N 2009, 'Western scrub jays conceal auditory information when competitors can hear but cannot see', *Biology Letters*, vol. 5, no. 5, pp. 583–5.

ACKNOWLEDGEMENTS

I thank my MUP editors and publishers, Lucy Davison, Cathy Smith, Collette Vella and Louise Adler, my agent Mary Cunnane, and my wife Penny for reading various versions of the text. Rob Day from the University of Melbourne introduced me to the ongoing Hawaiian malaria disaster, while Jim Kaufman of Cambridge University updated my understanding of chicken immunogenetics and reviewed what I wrote on that subject. At the earliest stages, colleagues from the School of Veterinary Science at the University of Melbourne and at Museum Victoria helped clarify my thinking and pointed me to specialists who could provide useful insights. Micheline Morgan kindly allowed me to present her previously unpublished, and delightful, poem. Jim Morgan and Alexis Beckett provided helpful discussion. Many who feature in the stories told here discussed their work with me and took the trouble to correct what I wrote. In this regard, I owe thanks to Gerry Swan, Ralph Doherty, Holly Earnest, Tom Monath, Hugh Reid, Ian Frazer, Rob Webster, Chuck Sherr, Allan Baker, Lynda Chambers and Nicky Clayton, both for their help and for doing the wonderful science that informs this volume. Frances Brodsky read the manuscript from the viewpoint of a research geneticist who is also a passionate birder. Only a little of this book is drawn directly from my own field, while a great deal was new to me. One of the real pleasures that comes from exploring a broad theme is learning about areas of innovative and imaginative science that have the potential to inform all of us. These are great stories that receive far too little attention. I hope that they will intrigue you as much as they fascinated me.

INDEX